SpringerBriefs in Physics

Series Editors

Balasubramanian Ananthanarayan, Centre for High Energy Physics, Indian Institute of Science, Bangalore, India

Egor Babaev, Department of Physics, Royal Institute of Technology, Stockholm, Sweden

Malcolm Bremer, H. H. Wills Physics Laboratory, University of Bristol, Bristol, UK

Xavier Calmet, Department of Physics and Astronomy, University of Sussex, Brighton, UK

Francesca Di Lodovico, Department of Physics, Queen Mary University of London, London, UK

Pablo D. Esquinazi, Institute for Experimental Physics II, University of Leipzig, Leipzig, Germany

Maarten Hoogerland, University of Auckland, Auckland, New Zealand

Eric Le Ru, School of Chemical and Physical Sciences, Victoria University of Wellington, Kelburn, New Zealand

Dario Narducci, University of Milano-Bicocca, Milan, Italy

James Overduin, Towson University, Towson, USA

Vesselin Petkov, Montreal, Canada

Stefan Theisen, Max-Planck-Institut für Gravitationsphysik, Golm, Germany

Charles H. T. Wang, Department of Physics, University of Aberdeen, Aberdeen, UK

James D. Wells, Department of Physics, University of Michigan, Ann Arbor, MI, USA

Andrew Whitaker, Department of Physics and Astronomy, Queen's University Belfast, Belfast, UK

SpringerBriefs in Physics are a series of slim high-quality publications encompassing the entire spectrum of physics. Manuscripts for SpringerBriefs in Physics will be evaluated by Springer and by members of the Editorial Board. Proposals and other communication should be sent to your Publishing Editors at Springer.

Featuring compact volumes of 50 to 125 pages (approximately 20,000–45,000 words), Briefs are shorter than a conventional book but longer than a journal article. Thus, Briefs serve as timely, concise tools for students, researchers, and professionals.

Typical texts for publication might include:

- A snapshot review of the current state of a hot or emerging field
- A concise introduction to core concepts that students must understand in order to make independent contributions
- An extended research report giving more details and discussion than is possible in a conventional journal article
- A manual describing underlying principles and best practices for an experimental technique
- An essay exploring new ideas within physics, related philosophical issues, or broader topics such as science and society

Briefs allow authors to present their ideas and readers to absorb them with minimal time investment. Briefs will be published as part of Springer's eBook collection, with millions of users worldwide. In addition, they will be available, just like other books, for individual print and electronic purchase. Briefs are characterized by fast, global electronic dissemination, straightforward publishing agreements, easy-to-use manuscript preparation and formatting guidelines, and expedited production schedules. We aim for publication 8–12 weeks after acceptance.

Aditya Vats · Varsha Banerjee · Sanjay Puri

Ferronematics and Living Liquid Crystals

Aditya Vats
Department of Physics
Indian Institute of Technology Delhi
New Delhi, Delhi, India

Varsha Banerjee
Department of Physics
Indian Institute of Technology Delhi
New Delhi, Delhi, India

Sanjay Puri
School of Physical Sciences
Jawaharlal Nehru University
New Delhi, Delhi, India

ISSN 2191-5423   ISSN 2191-5431   (electronic)
SpringerBriefs in Physics
ISBN 978-3-031-87798-8   ISBN 978-3-031-87799-5   (eBook)
https://doi.org/10.1007/978-3-031-87799-5

© The Author(s), under exclusive license to Springer Nature Switzerland AG 2025

This work is subject to copyright. All rights are solely and exclusively licensed by the Publisher, whether the whole or part of the material is concerned, specifically the rights of translation, reprinting, reuse of illustrations, recitation, broadcasting, reproduction on microfilms or in any other physical way, and transmission or information storage and retrieval, electronic adaptation, computer software, or by similar or dissimilar methodology now known or hereafter developed.

The use of general descriptive names, registered names, trademarks, service marks, etc. in this publication does not imply, even in the absence of a specific statement, that such names are exempt from the relevant protective laws and regulations and therefore free for general use.

The publisher, the authors and the editors are safe to assume that the advice and information in this book are believed to be true and accurate at the date of publication. Neither the publisher nor the authors or the editors give a warranty, expressed or implied, with respect to the material contained herein or for any errors or omissions that may have been made. The publisher remains neutral with regard to jurisdictional claims in published maps and institutional affiliations.

This Springer imprint is published by the registered company Springer Nature Switzerland AG
The registered company address is: Gewerbestrasse 11, 6330 Cham, Switzerland

If disposing of this product, please recycle the paper.

To our parents

Preface

The book stems from the work done for the Ph.D. thesis of Aditya at the Indian Institute of Technology Delhi, New Delhi, India. It deals with two contemporary soft matter systems: *ferronematics* (FNs) and *living liquid crystals* (LLCs). Both fall into the broader category of nematic liquid crystals (NLCs) with inclusions. They are being extensively studied through experiments because of the emergence of remarkable properties that are not present in their pure counterparts. Novel theoretical modeling and extensive numerical simulations are used to comprehensively understand them. The overarching aim of the book is to offer theoretical techniques appropriate for studying these systems.

The book unfolds systematically, beginning with an introduction to FNs and LLCs and the current status of their research. The second chapter provides the theoretical framework necessary to explore the dynamics of FNs and LLCs. The framework integrates the well-established coarse-grained formalism for NLCs with phenomenological models for magnetic or active particles. The coupling between the components is constructed by taking cues from experimental observations. The coarse-grained approach has enabled the study of large-scale behavior to unravel novel states and their dynamical evolution. Special emphasis has been placed on understanding the role of coupling in textures, defects, and phase transitions.

The subsequent chapters explore the important nonequilibrium phenomenon of *phase ordering dynamics* in these systems. For FNs, the dynamics of two-component coupling is examined, revealing novel behavior such as *slaved coarsening*. The emergence of *biaxiality* through magneto-nematic coupling is studied in detail, yielding theoretical predictions that align with experimental observations. For LLCs, a new model is proposed to study pattern dynamics—this model uncovers novel steady states such as *chimeras* and *solitons*. These findings have significant implications, including the possibility of designing self-healing materials and developing advanced systems for targeted transport and energy harvesting.

This book addresses fundamental aspects of the framework for two-component soft matter systems. The versatility of the approach makes it applicable to polymers, colloids, biological systems, and even the newly discovered ferroelectric nematics. The inclusion of hydrodynamic effects and non-reciprocal interactions represents an interesting future direction with the promise of expanding the scope of this work further. This book aims to inspire fundamental research and innovative applications by presenting a unified theoretical framework and predictive insights into complex soft matter systems. It is designed for researchers in materials science, condensed matter physics, and related fields, as well as graduate students interested in coarse-grained modeling for nonequilibrium studies. It is hoped that the novel methodology presented here will pave the way for future explorations in the fascinating world of soft matter.

Before concluding, we extend our heartfelt gratitude to everyone who contributed to this project. The authors sincerely thank Pradeep Kumar Yadav who collaborated with us on some of these problems. We also thank other members of our lab: Konark Bisht, Nishant Birdi, Anuj Kumar Singh, Parbati Saha, and Ritik Rajak. They have been instrumental in cultivating a healthy environment for scientific discussion and inquiry. We gratefully acknowledge the following Indian funding agencies for financial support during the course of our research: *Council for Scientific and Industrial Research, Department of Science and Technology*, and *Anusandhan National Research Foundation*. Finally, we thank our families for their unwavering support and patience.

New Delhi, India
Aditya Vats
Sanjay Puri
Varsha Banerjee

Competing Interests The authors have no competing interests to declare that are relevant to the content of this manuscript.

Contents

1 **Introduction** 1
   1.1 Ferronematics 3
   1.2 Living Liquid Crystals 7
   1.3 Outline of the Book 10
   References 11

2 **Theoretical Background** 15
   2.1 Free Energy Models 15
      2.1.1 Nematic Free Energy 16
      2.1.2 Ginzburg–Landau Free Energy for Magnetic System 18
      2.1.3 Free Energy for Active Particles 19
   2.2 Dynamical Equations 20
      2.2.1 Time-Dependent Ginzburg–Landau Equations 20
      2.2.2 Toner–Tu Equations for Active Particles 21
   2.3 Numerical Techniques Used to Solve Dynamical Equations 22
   References 23

3 **Phase Ordering Dynamics in Ferronematics** 25
   3.1 Introduction 25
   3.2 Theoretical Framework 27
      3.2.1 Ginzburg–Landau Free Energy 27
      3.2.2 Time-Dependent Ginzburg–Landau Equations 27
      3.2.3 Fixed-Point Solutions for $d = 2$ FNs 30
      3.2.4 Morphology Characterization 32
   3.3 Detailed Numerical Results 34
      3.3.1 Domain Growth in $d = 2$ Ferronematics 34
      3.3.2 Domain Growth in $d = 3$ Ferronematics 41
   3.4 Summary and Discussion 45
   References 47

4 Emergence of Biaxial Order in Ferronematics 49
- 4.1 Introduction 49
- 4.2 Theoretical Framework 51
  - 4.2.1 Coarse-Grained Free Energy for Ferronematics 51
  - 4.2.2 Time-Dependent Ginzburg–Landau Equations 52
- 4.3 Results 53
  - 4.3.1 Ordering Kinetics 54
  - 4.3.2 Emergence of Biaxiality 54
- 4.4 Summary and Discussion 58
- References 59

5 Symbiotic Dynamics in Living Liquid Crystals 61
- 5.1 Introduction 61
- 5.2 Theoretical Framework 63
  - 5.2.1 Model 63
  - 5.2.2 Fixed-Point Solutions and Linear Stability Analysis 66
- 5.3 Results for LLCs in Bulk 68
- 5.4 Surface-Directed Dynamics in LLCs 72
- 5.5 Summary and Discussion 78
- Appendix: Movies Showing Steady States of LLCs 80
- References 80

6 Conclusion and Perspectives 83
- References 85

Appendix A: Stationary Solutions 87

Abbreviations

BNLC	Biaxial Nematic Liquid Crystal
BPT	Bray-Puri-Toyoki
CH	Cahn-Hilliard
FN	Ferronematic
FP	Fixed-Point
GL	Ginzburg-Landau
LAC	Lifshitz-Allen-Cahn
LC	Liquid Crystal
LCD	Liquid Crystal Display
LdG	Landau-de Gennes
LLC	Living Liquid Crystal
LRO	Long-Range Order
LS	Lifshitz-Slyozov
MNP	Magnetic Nanoparticle
NLC	Nematic Liquid Crystal
SDM	Sub-domain Morphology
TDGL	Time-Dependent Ginzburg-Landau
TT	Toner-Tu

Chapter 1
Introduction

Abstract This chapter introduces two contemporary systems in soft matter physics: *ferronematics* (FNs) and *living liquid crystals* (LLCs). FNs are nematic liquid crystals that incorporate magnetic nanoparticles, whereas LLCs consist of nematics interspersed with active, self-propelled agents. The chapter provides a comprehensive overview of the fundamental principles that underlie these systems, emphasizing the interplay between orientational order and dynamic response. The discussion focuses on the experimental observations and distinctive features of FNs and LLCs. It sets the groundwork for exploring the rich dynamical behavior and emergent properties of these complex materials.

The study of materials has undergone significant transformations as a result of the innovative practice of doping. This involves the deliberate addition of specific elements or compounds (*inclusions*) to a base material, fundamentally altering its inherent properties to improve performance and functionality. The roots of these methods extend back to ancient alchemists, who discovered that inclusions could dramatically change the characteristics of metals, laying the groundwork for more systematic advancements. A notable historical example is the addition of carbon to iron to produce steel, which revolutionized construction and toolmaking by creating stronger, more resilient materials. In the modern era, doping has become crucial in semiconductor technology, where the introduction of tiny impurities in silicon crystals allows precise control of electrical properties. This has been vital for the development of contemporary electronics, such as transistors, diodes, and integrated circuits. The ability to fine-tune conduction in semiconductors has driven the digital revolution and spurred progress in telecommunications, computing, and renewable energy. In addition to altering electrical conductivity in metals and semiconductors, strategic inclusions enhance the mechanical properties, thermal stability, and chemical resistance of ceramics. Polymer additives, on the other hand, increase flexibility, strength, and durability, which has led to innovations in medicine, aerospace, and consumer goods. From ancient alchemists' experiments to today's sophisticated techniques, inclusions and doping have propelled material science forward and driven technological progress across diverse industries.

In recent years, soft-matter systems have acquired immense importance in various fields of science and technology. Their significance stems from the fact that they bridge the gap between traditional solids and liquids, offering a wide range of functionalities and applications. In particular, inclusion in soft matter systems such as colloidal glasses, polymers, emulsions, granular materials, LCs, and gels has significantly improved their mechanical, thermal, and rheological properties [1–4]. Some examples are metallic ions that impart a distinct color to colloidal glasses [5–7], a blend of two polymers that shows increased flexibility [8], and nanoparticles in biological systems that enhance targeted drug delivery and imaging techniques [9, 10]. In addition to a wide array of applications, inclusions provide valuable insight into the structure-property relationship, phase transitions, and the self-assembly process. Dopant particles alter the energy landscape of the material and shed light on the fundamental thermodynamic principle governing the system. Therefore, the inclusion of specific elements or compounds in soft matter systems not only imparts unique properties and expands their applications, but also provides a richer understanding of the fundamental principles of material science. Extensive theoretical research precedes the practical application of these doped soft matter compounds with the aim of modeling and quantifying the induced modifications. In this pursuit, the statistical mechanics framework has found widespread application, offering insightful perspectives into the behavior, properties, and characteristics of materials driven by inclusions. Using this framework, researchers have gained a profound understanding of the intricate interplay between impurities and host materials, paving the way for further advancements in diverse domains.

This book studies the intricate world of soft matter physics, specifically focusing on two contemporary complex systems: (i) *ferronematics* and (ii) *living liquid crystals*. Ferronematics (FNs) are a composite system in which nematic liquid crystals (NLCs) are doped with magnetic nanoparticles (MNPs), whereas living liquid crystals (LLCs) are an amalgamation of living particles such as bacteria in LCs. There are exciting experiments with both FNs and LLCs, but models and methodologies for studying them are few and far between. The main purpose of this book is to fill this lacuna by providing a theoretical framework for studying equilibrium and nonequilibrium properties in these coupled systems. The focus is on exploring the effect of inclusions on textures, defects, flows, and their manipulation in different dimensions. In the process, novel phases with potential technological applications are identified. This chapter introduces the research problems and is organized as follows. In Sect. 1.1, an introduction to the FN suspensions is provided, explaining how the inclusion of magnetic particles in NLCs makes them useful for novel magneto-optic effects. Section 1.2 introduces LLCs and discusses some of their experimental observations. An overview of the book is presented in Sect. 1.3.

1.1 Ferronematics

LCs are mesomorphic states between ordinary liquids and solids. The constituent molecules translate freely as in a liquid while exhibiting the long-range orientation order seen in solids [11–14]. They exist in a variety of phases, namely nematic, smectic, cholesteric, columnar, twist-bend, splay-bend, etc. [15]. The simplest LCs among them are NLCs, where the molecules are often rod-like or disc-shaped. They typically undergo a phase transition from isotropic to nematic state when quenched below a critical temperature T_c^N [11, 12]. Figure 1.1 shows schematically the effect of temperature on the order of rod-shaped molecules. They tend to align parallel because this is energetically favorable, but are randomized by entropy. At higher temperatures, entropy prevails and the result is a disordered isotropic state. However, at lower temperatures, energy dominates, leading to a fully aligned crystal state. For intermediate temperatures, an LC state with orientational order but no positional order is favored. The special class of NLC molecules exhibits strong anisotropy and has a natural tendency to align parallel to each other. This preferred direction of alignment is described by a sign-invariant unit vector called *director* **n** (see schematic in Fig. 1.2). Within the nematic phase, the system can exist in the *uniaxial* or *biaxial* state [16–18], as shown in Fig. 1.2. The order is uniaxial if the molecular alignment is only about **n**. Biaxial nematic liquid crystals (BNLCs) have an additional distinguished (secondary) director **k** ($\perp$ to **n**) for orientational ordering. In contrast to a uniaxial state where there is a rotational symmetry about a single direction, BNLCs have three special directions of reflection symmetry; see the schematic in Fig. 1.2.

The mechanical, optical and diffusive properties of NLCs show a strong directional dependence [19]. They exhibit unique dielectric and diamagnetic characteristics that make them suitable for a range of applications. Their utility in modern liquid crystal displays (LCDs) and optical imaging is derived from the rapid reorientation of **n** achievable within milliseconds under low electric fields (1–2 mV) [20].

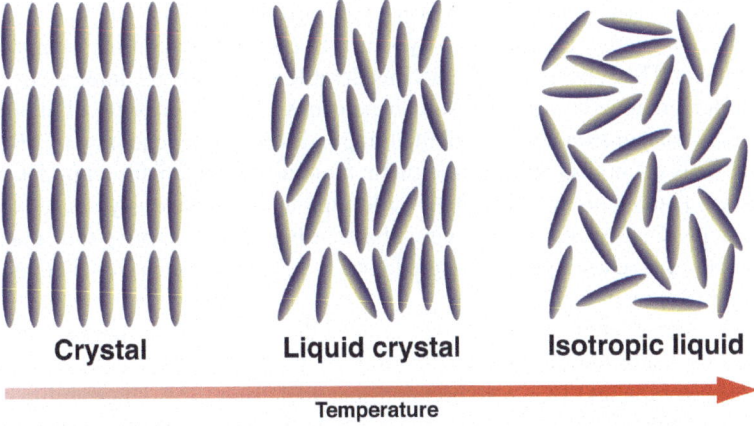

Fig. 1.1 Schematic diagram depicting the effect of temperature on order in rod-shaped molecules

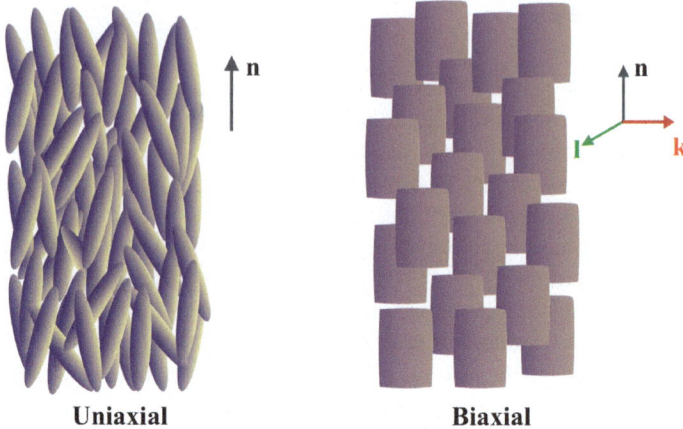

Fig. 1.2 Schematic diagram of the rod-shaped and book-shaped molecules exhibiting the uniaxial and biaxial ordering

This electric sensitivity allows for precise control over the optical properties of the products. However, they are not responsive to magnetic fields due to their low magnetic susceptibility $\sim 10^{-6}$ (in SI units). Consequently, NLCs require strong magnetic fields—around 300 mT—to achieve actuation, which has limited the scope of their application [21].

A natural question that arises then is whether adding a small quantity of magnetic material into NLCs could enhance their magnetic sensitivity, thereby enabling magneto-optic response alongside the traditional electro-optic actuation. In a pioneering 1970 study, Brochard and de Gennes proposed this idea by suggesting the addition of a minimal amount of ferromagnetic particles (approximately 0.01% by volume) to NLCs, aiming to significantly boost their magnetic responsiveness [22]. Their concept is illustrated schematically in Fig. 1.3. Brochard and de Gennes theorized that magnetic grains anchored by the surrounding NLC molecules could increase the magnetic susceptibility by several orders of magnitude compared to pure nematics. This concept spurred considerable experimental interest in creating stable ferromagnetic liquid crystal suspensions [23–34]. In spite of significant efforts, stable suspensions proved challenging because of the tendency of MNPs to aggregate. This occurs because the magnetic moments align in parallel under strong interactions. This favors ferromagnetic ordering but simultaneously causes clumping of particles. Consequently, achieving a balance between promoting ferromagnetic alignment and preventing particle aggregation remained a primary challenge in the practical development of FNs.

In 2013, Mertelj et al. successfully produced the first stable suspension of ferromagnetic LCs using barium hexaferrite (*BaHF*) magnetic nanoplatelets suspended in pentylcyano-biphenyl LCs with homeotropic anchoring [35]. The stability of this suspension is attributed to a subtle interaction between the platelet shape and the

1.1 Ferronematics

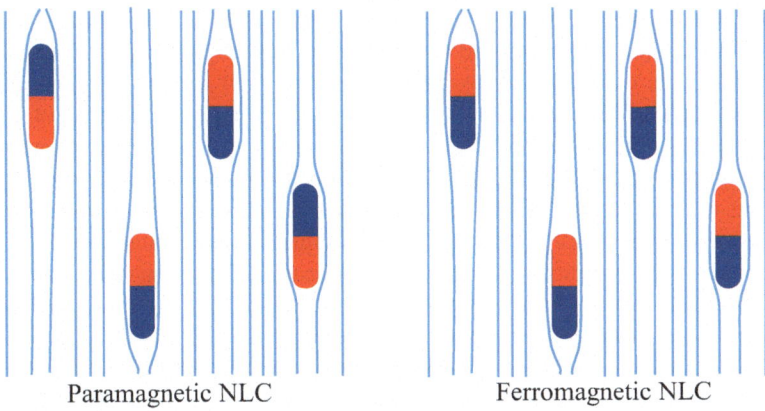

Fig. 1.3 Schematics depicting paramagnetic and ferromagnetic nematic suspension. The red and blue halves represent the north and south poles of the magnetic grains

anchoring of surrounding LC molecules. When a particle is immersed in a liquid crystal, it disrupts the uniform director field, causing an elastic deformation that can be described by a series of multipolar terms. In cases where no external torque acts on the particle, the uniaxial symmetry of the LC eliminates only the monopole term. The contributions of the rest of the terms (dipolar, quadrupolar, etc.) are influenced by the shape of the particle and the anchoring conditions. They induce a *topological charge* such as ± 1 or $\pm 1/2$. For example, dipolar interactions generally produce point defects in the LC, while quadrupolar interactions can create more complex structures such as disclination loops or a characteristic 'Saturn ring' around the particle. In the case of *BaHF* disc-shaped particles, Mertelj et al. observed a quadrupolar symmetry in the surrounding LC director field, introducing a net topological charge of $+1$. This charge requires compensation through the formation of an LC defect with an opposite charge, specifically a Saturn ring with a charge of -1 [36]. The interaction between the Saturn ring defect and the LC itself, both carrying identical topological charges, could counteract the magnetic dipole-dipole attraction between particles, maintaining stability. This approach provided a pathway for mitigated aggregation while preserving the ferromagnetic behavior essential for functional LC suspensions.

There are two experimental protocols that are employed to obtain stable FNs. In the first method, the MNP-NLC mixture is quenched from the (disordered) isotropic phase to the (ordered) nematic phase in the presence of an external magnetic field applied along the director's direction. This procedure yields the FN domain with *aligned* magnetic moments, and the resulting suspensions are termed *ferromagnetic* NLCs [36]. In this protocol, it is essential to incorporate an additional field-dependent term in the formalism to represent the coupling between the applied field and the magnetic orientations. In the second method, the MNP-NLC suspension is quenched without an external magnetic field. The moments align along **n** or $-$**n** due to the magneto-nematic coupling and yield a vanishing macroscopic magnetization [37, 38].

FNs are receiving increasing interest from both academia and industry [39–49]. They offer a unique opportunity to study the effect of magnetic inclusions on fundamental aspects of LCs, such as phase transition, self-assembly, rheology, and optical properties. This behavior is fundamentally different from that of typical ferromagnetic solids, where the particles at defect sites cannot move. However, in a ferromagnetic liquid, the fluidity allows magnetic moments, domain walls, and impurities to move freely. This mobility dramatically alters the processes of domain formation and domain wall dynamics. These unique properties make ferromagnetic liquids or FNs highly attractive for a range of applications. In recent years, their strong magneto-optic effect has been successfully demonstrated for magnetic field visualization [50], highlighting their potential in advanced sensing and display technologies. Unlike conventional NLCs, where electrical switching is restricted by electrode placement, FNs can be switched magnetically, offering noncontact and multidirectional control. This capability opens up new possibilities for remote manipulation of LC devices, leading to more versatile and dynamic pixel control in display technologies and enhanced light modulation in optical communication systems. In addition, the tunable polarity of their magnetic response allows for precise control over material behavior, making them ideal for advanced sensors and actuators. Therefore, FNs give access to a coupled system with exotic morphologies, interesting defect structures, and a new theoretical formulation that is not accessible in an uncoupled system. In addition, there have been active efforts to utilize magneto-mechanical and magneto-optic effects for applications in photonics [47], optical switches [40], complex fluids [41, 42], and even particle physics and cosmology [44].

Following the creation of stable FN suspensions, efforts have largely focused on exploring how the coupling between nematic and magnetic components affects both equilibrium and non-equilibrium properties. Several experimental studies have offered fascinating insights. For example, Shuai et al. demonstrated that magneto-nematic coupling could spontaneously generate flux closure loops sensitive to fields as weak as the Earth's magnetic field [46]. In a separate study, Mertelj and Lisjak cooled a ferronematic droplet from the isotropic phase to the nematic phase under a field aligned with the nematic director **n** [36]. By applying magneto-optic techniques, they observed the formation of bubbles or small domains with magnetization parallel to or antiparallel to the field, which evolved over time. Liu et al. conducted another seminal experiment, finding that by embedding magnetic nanoparticles in the NLC medium, a biaxial order could be achieved [40]. Their approach capitalized on the distinct length scales of dipolar and magneto-nematic interactions, producing an equilibrium state where the magnetic moments of the nanoparticles aligned at an angle to the nematic director **n**, thereby introducing a new order direction in the perpendicular plane. The confirmation of the biaxial order came from the system's absorption spectrum and magnetic hysteresis profiles. These advances broaden the scope of potential FN applications, highlighting the need for theoretical insights to guide these explorations.

The first part of this book focuses on understanding the effects of two-component coupling in FNs. In this context, non-equilibrium studies of domain growth or coarsening after a deep temperature quench are particularly insightful. In a typical

1.2 Living Liquid Crystals

coarsening experiment, the system is immediately cooled from a disordered phase ($T > T_c$) to an ordered phase ($T < T_c$). The highly non-equilibrium system then evolves toward equilibrium by gradually eliminating defects [51, 52]. This evolution is marked by the formation and growth of ordered domains and provides deep insights into the complex energy landscape and relaxation time scales of the system. Several key questions for FNs, which have *two* interacting order parameters: (i) What can coarsening experiments reveal about FNs' behavior? (ii) How does the system behave if only one of the two components reaches an ordered state? (iii) What types of defect structures are present and how do these vary across dimensions? (iv) Could this intricate system reveal new phases or textures? The book aims to construct a theoretical framework to address these questions and ultimately uncovers complex textures and new phases in this coupled system.

1.2 Living Liquid Crystals

The second part of the book deals with active inclusions in NLCs. Active matter (AM) is an assembly of interacting particles that converts energy from the environment into mechanical motion for self-propulsion. These exhibit coherent dynamic motion on a scale much larger than individual units. Active particles are typically elongated, with their direction of self-propulsion determined by the intrinsic anisotropy rather than by an external field. As a result, orientational order is a prevalent theme in the study of active matter. The constituent particles can be biological or synthetic and range from micrometers to several meters. Some prototypical examples are polar gels, bacterial suspensions, microtubule bundles, bird flocks, fish shoals, vibrating granular gases, etc.[53–69]. Active particles generally react to the motion of neighboring individuals that leads to their synchronized motion; see the schematic diagram in Fig. 1.4 for such a representation.

A suspension of microscopic particles such as bacteria or synthetic swimmers in the isotropic medium has been the most characterized AM system. They exhibit turbulent motion because thermal noise and steric collisions cause the random tumbling or flipping of individual particles [67, 70–73]. Consequently, several novel features emerge, such as reduction in effective viscosity, enhanced self-diffusion, and active clustering that are quite different from the properties of equilibrium colloidal suspensions [60, 62, 67–71]. Further interesting opportunities to control and manipulate AM arise if an anisotropic material is used as the suspending medium. It offers long-ranged order and directional dependence for the transportation of the active particles. In this context, a topical system is *living liquid crystals* where living (active) particles are introduced into NLCs [74–81]. It is essential to differentiate liquid crystal colloids from extensively studied active nematics [69, 82]. These comprise of rod-like active particles that self-organize into large-scale structures with orientational order and generate self-sustained flows. Examples include kinesin molecular motors and microtubules that operate in the presence of adenosine triphosphate (ATP) and a crowding agent such as polyethylene glycol (PEG), as well as vibrating granular

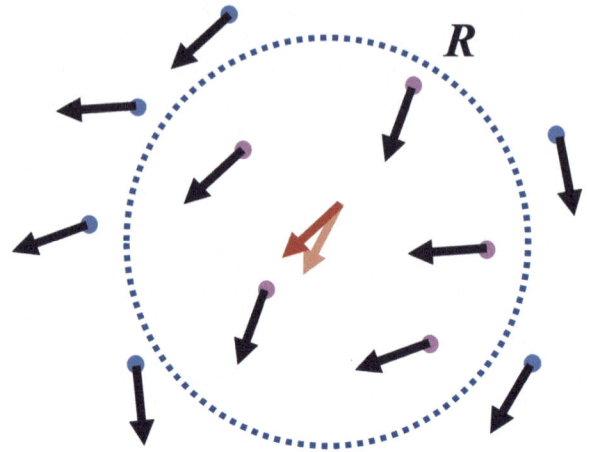

Fig. 1.4 Schematic diagram depicting the alignment of active particles toward the average direction of their neighbours

rods. In molecular motors for kinesin, active kinesin clusters attach to pairs of microtubules with opposite polarity, causing them to slide relative to each other, leading to a displacement of their centers of mass. This relative sliding produces "active stresses" in the surrounding medium, which drives the self-organization process. In contrast, when bacteria swim in a liquid crystal medium, the energy input is individual with movement powered by the rotation of their flagella. Furthermore, there is no direct mechanical coupling between the bacteria's orientation and the liquid crystal molecules; instead, the interaction occurs through hydrodynamic and elastic forces. Therefore, the applicability of generic phenomenological models of active nematics to LLCs is somewhat restricted. To accurately describe LLCs, more detailed and specialized theoretical approaches need to be developed.

Contemporary studies on LCs have focused on suspensions with low concentrations of rod-shaped bacteria, such as *Bacillus subtilis*, in non-toxic NLCs confined to quasi-2D geometries. These configurations reveal fascinating phenomena unique to LLCs and never observed in traditional Newtonian fluids [74, 76, 78, 83]. The bacterial flagella act as probes, interacting with NLCs at the nanoscale and generating emergent textures that span hundreds of micrometers. Topological defects in NLCs play a pivotal role in bacterial dynamics. In defect-free regions, bacteria align with the local nematic director, while T-shaped defects with a topological charge $+1/2$ attract bacteria and Y-shaped defects with a charge $-1/2$ repel them [78]. The trajectories around these defects are shown in Fig. 1.5. These observations likely extend to other self-propelled particles, including synthetic swimmers, provided the concentration remains low. LLCs are anticipated to bridge the properties of active and passive materials to enable advanced microfluidic devices that transport fluids without mechanical pumping, mimic cellular motion, and develop nanotechnologies for targeted delivery, sensing, and biomedical uses.

1.2 Living Liquid Crystals

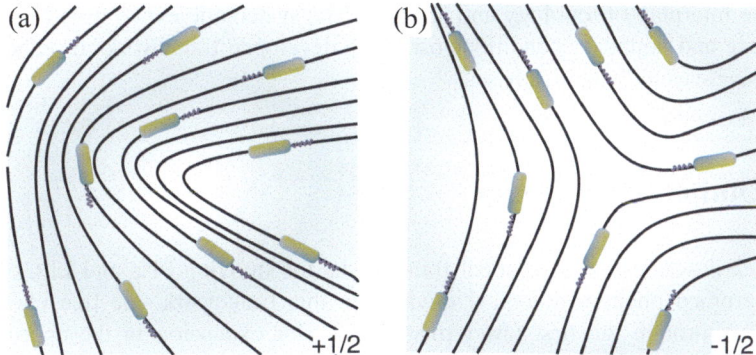

Fig. 1.5 Schematic diagram depicting the bacterial motion around **a** +1/2 and **b** −1/2 nematic defects

An important advancement in LLC research involves developing models that enable theoretical and experimental synergy to unlock their potential applications. One of the first attempts in this direction is by Genkin et al. [78], who devised continuum equations for the observed behavior of rod-shaped bacteria in NLCs. Based on experimental observations, the model makes key assumptions: (i) the volume fraction of the bacteria is low enough to avoid significantly affecting the NLC's properties; (ii) bacteria align with the local nematic director rapidly compared to the collective behavioral timescale; (iii) the interactions of the bacteria are apolar, allowing them to glide without collision across the quasi-2D space. The model applies the Beris-Edwards equations for the tensor order parameter $\mathbf{Q}(\mathbf{r}, t)$ and velocity $\mathbf{u}(\mathbf{r}, t)$, representing the NLC environment. Bacterial transport is then governed by coupled advection-diffusion equations for concentrations of bacteria moving parallel (c^+) and antiparallel (c^-) to the director $\mathbf{n}$ [78, 84, 85]. This model effectively reproduces the experimentally observed bacterial accumulation and repulsion at defect sites in the dilute regime, where bacterial interactions are minimal. Although the work offers valuable insights in the dilute regime, where bacterial interactions are minimal, it leaves the dense limit largely unexplored. The experiments by Zhou et al. [74] highlighted the rich phenomenology of LLCs in this dense regime, showing the need to move beyond the dilute limit for their description.

This book presents coarse-grained models to study LLC dynamics, capturing pattern formation with all three interaction types: active matter (AM-AM), liquid crystal (LC-LC), and active-nematic (AM-LC) influencing the behavior. The models are guided by bacteria-in-NLC experiments and broadly apply to any synthetic swimmer in anisotropic media. The framework also incorporates the largely unexplored dense limit essential for many active-matter applications. Key questions that are addressed in LLC pattern formation include: (i) How does coupling impact defect structures? (ii) Can nematic textures remain stable in the presence of active particles? (iii) Can active trajectories be directed by modifying topological defects in the nematic matrix? (iv) How do boundary conditions influence the structures arising from the interplay of topology and activity? The systematic exploration of parameter space and boundary conditions using the developed frameworks reveals novel morphologies and dynamic phases.

1.3 Outline of the Book

Chapter 2 describes the theoretical frameworks for studying FNs and LLCs using coarse-grained phenomenological models. In this framework, the free energy is expressed through the first few terms of a Taylor expansion in the appropriate order parameter, capturing phase transitions, critical phenomena, and other emergent properties. Such models have successfully bridged experimental observations and are computationally inexpensive compared to precise microscopic models that require detailed knowledge of molecular shapes and intermolecular potentials. First, the Landau–de Gennes free energy model for NLCs, the Ginzburg–Landau free energy for the magnetic component, and the Toner–Tu model for AM are described. These provide the necessary background to formulate the coarse-grained free energy description for composite FNs and LLCs. The free energy minimum is reached by defining the appropriate equations of motion. In addition, a pedagogical overview is provided for obtaining fixed point solutions and performing a linear stability analysis to predict stable (and unstable) solutions at equilibrium. Finally, the chapter concludes with a numerical technique to efficiently solve the dynamical equations, which are anticipated to be more challenging for the FN and LLC systems.

Chapter 3 is dedicated to exploring the effects of two-component coupling in FNs [86, 87]. Using a coarse-grained Landau–de Gennes free energy approach, the ordering dynamics is analyzed in both $d = 2$ and $d = 3$ dimensions. Through extensive analytical and numerical investigations, various insights into phase-transition kinetics are presented. The ordering behavior in response to thermal quenches is observed to follow key patterns: (i) For shallow quenches ($T_c^N < T < T_c^M$, with N and M representing nematic and magnetic transitions), the magnetically ordered component can *enslave* the nematic component, aligning their domains. This dependency is also observed under the reverse conditions where $T_c^M < T < T_c^N$. (ii) When the coupling is asymmetric, distinct subdomain morphologies (SDMs) emerge, marked by interfacial defects. (iii) For these SDMs, the structure factor $S(k)$ shows *Porod decay*, $S(k) \sim k^{-(d+1)}$, characteristic of scattering from sharp interfaces in d-dimensional space.

In Chap. 4, the discussion progresses to the emergence of biaxial order within FNs. Experimentally, the realization of a stable biaxial phase has been challenging due to the fragility of secondary molecular alignments under thermal fluctuations. A recent breakthrough by Liu et al. has demonstrated stable biaxiality in FNs [40]. This chapter provides a theoretical framework to understand the biaxial order in FNs, emphasizing how magneto-nematic coupling introduces biaxiality. By properly designing the coupling term, the system can achieve a stable state with the perpendicular alignment of **n** and **M**. This coupling adds an additional axis of order (**k**) perpendicular to **n**, at minimal energetic cost, leading to biaxial order. The strength of the biaxiality is quantitatively analyzed as a function of coupling strength, with results well supported by fixed-point analytical solutions.

Chapter 5 introduces a phenomenological model for the dynamics of LLCs. This approach combines the Toner–Tu model for active matter with the Landau–de Gennes free energy for NLCs along with an experimentally motivated coupling term [78]. Given that the Landau–de Gennes model lacks hydrodynamic terms, director dynamics is imparted by time-dependent Ginzburg–Landau (TDGL) equations with purely relaxational dynamics [51, 52]. Through simulations, two novel steady states emerge in LLC systems: (i) *Chimera states* or bands of high orientational order (in both AM and NLCs) within a disordered background of active matter and isotropic NLCs, which propagate through the system at the speed v_0 of active particles [88, 89]; (ii) Localized solitons, rare in dimensions >1, that exhibit high orientational order in both AM and NLCs and move at speed v_0. The phase diagram is determined analytically by evaluating fixed points of the dynamical equations and performing a linear stability analysis. Director patterns and active flow are tuned using relevant boundary conditions, revealing dynamic steady states governed by the coupling strength. Notable outcomes include persistent vortex states within microfluidic wells, tailored morphologies, and dynamic states with potential applications in drug delivery and chaotic active particle control.

Finally, in Chap. 6, the book concludes with a discussion of some open problems and the possibility of adaptation of the frameworks developed here to other two-component systems in different physical settings.

References

1. S.R. Nagel, Rev. Mod. Phys. **89**, 025002 (2017)
2. H. Wu, H. Friedrich, J.P. Ptterson, N. Sommerdijk, N. De Jonge, Adv. Mater. **32**, 2001582 (2020)
3. L.S. Hirst, *Fundamentals of Soft Matter Science* (CRC Press, 2019)
4. A. Fernandez-Nieves, A.M. Puertas, *Fluids, Colloids and Soft Materials: An Introduction to Soft Matter Physics* (Wiley, 2016)
5. P.J. Lu, D.A. Weitz, Annu. Rev. Condens. Matter Phys. **4**, 217 (2013)
6. J.B. Kim, S.Y. Lee, J.M. Lee, S.H. Kim, A.C.S. Appl. Mater. Interfaces **11**, 14485 (2019)
7. K. Ueno, Y. Sano, A. Inaba, M. Kondoh, M. Watanabe, J. Phys. Chem. B **114**, 13095 (2010)
8. A. Sionkowska, Prog. Polym. Sci. **36**, 1254 (2011)
9. M. De, P.S. Ghosh, V.M. Rotello, Adv. Mater. **20**, 4225 (2008)
10. O.V. Salata, J. Nanobiotechnology **2**, 1 (2004)
11. P.G. de Gennes, J. Prost, *The Physics of Liquid Crystals* (Oxford University, Oxford, 1995)
12. E. Priestly, *Introduction to Liquid Crystals* (Springer Science, 2012)
13. M.J. Stephen, J.P. Straley, Rev. Mod. Phys. **46**, 617 (1974)
14. O.D. Lavrentovich, M. Kleman, *Cholesteric Liquid Crystals: Defects and Topology* (Springer, 2001)
15. P.J. Collings, J.W. Goodby, *Introduction to Liquid Crystals: Chemistry and Physics* (CRC Press, 2019)
16. C. Tsakonas, A.J. Davidson, C.V. Brown, N.J. Mottram, Appl. Phys. Lett. **90**, 111913 (2007)

17. G.R. Luckhurst, T.J. Sluckin, *Biaxial Nematic Liquid Crystals: Theory, Simulation and Experiment* (Wiley, 2015)
18. C. Tschierske, D.J. Photinos, J. Mater. Chem. **20**, 4263 (2010)
19. M.J. Stephen, J.P. Straley, Rev. Mod. Phys. **46**, 617 (1974)
20. H. Chen, J. Lee, B. Lin, S. Chen, S. Wu, Light Sci. Appl. **7**, 17168 (2018)
21. P.G. de Gennes, Mol. Cryst. Liq. Cryst. **7**, 325 (1969)
22. F. Brochard, P.G. de Gennes, J. Phys. **31**, 691 (1970)
23. J. Rault, P.E. Cladis, J.P. Burger, Phys. Lett. A **32**, 199 (1970)
24. S.H. Chen, N.M. Amer, Phys. Rev. Lett. **51**, 2298 (1983)
25. T. Kroin, A.M.F. Neto, Phys. Rev. A **36**, 2987 (1987)
26. S.V. Burylov, Y.L. Raikher, J. Magn, Magn. Mater. **122**, 62 (1993)
27. S.V. Burylov, Y.L. Raikher, Mol. Cryst. Liq. Cryst. **258**, 107 (1995)
28. M. Koneracka, V. Kellnerova, P. Kopčanskỳ, T. Kuczynski, J. Magn, Magn. Mater. **140**, 1455 (1995)
29. A.N. Zakhlevnykh, P.A. Sosnin, J. Magn, Magn. Mater. **146**, 103 (1995)
30. E. Jarkova, H. Pleiner, H.W. Müller, H.R. Brand, Jour. Chem. Phys. **118**, 2422 (2003)
31. N. Podoliak, O. Buchnev, O. Buluy, G. D'Alessandro, M. Kaczmarek, Y. Reznikov, T.J. Sluckin, Soft Matter **7**, 4742 (2011)
32. O. Buluy, S. Nepijko, V. Reshetnyak, E. Ouskova, V. Zadorozhnii, A. Leonhardt, M. Ritschel, G. Schönhense, Y. Reznikov, Soft Matter **7**, 644 (2011)
33. S.V. Burylov, A.N. Zakhlevnykh, Phys. Rev. E **88**, 012511 (2013)
34. N. Tomašovičová, M. Timko, Z. Mitróová, M. Koneracká, M. Rajňak, N. Eber, T. Tóth-Katona, X. Chaud, J. Jadzyn, P. Kopčanskỳ, Phys. Rev. E **87**, 014501 (2013)
35. A. Mertelj, D. Lisjak, M. Drofenik, M. Čopič, Nature **504**, 237 (2013)
36. A. Mertelj, D. Lisjak, Liq. Cryst. Rev. **5**, 1 (2017)
37. D.A. Petrov, A.N. Zakhlevnykh, Mol. Cryst. Liq. Cryst. **557**, 60 (2012)
38. A.N. Zakhlevnykh, D.A. Petrov, J. Magn. Magn. Mater. **401**, 188 (2016)
39. A. Mertelj, N. Osterman, D. Lisjak, M. Čopič, Soft Matter **10**, 9065 (2014)
40. Q. Liu, P.J. Ackerman, T.C. Lubensky, I.I. Smalyukh, Proc. Nat. Acad. Sci. **113**, 10479 (2016)
41. T. Potisk, D. Svenšek, H.R. Brand, H. Pleiner, H. Pleiner, D. Lisjak, N. Osterman, A. Mertelj, Phys. Rev. Lett. **119**, 097802 (2017)
42. T. Potisk, A. Mertelj, N. Sebastián, N. Osterman, D. Lisjak, H.R. Brand, H. Pleiner, D. Svenšek, Phys. Rev. E **97**, 012701 (2018)
43. Q. Zhang, P.J. Ackerman, Q. Liu, I.I. Smalyukh, Phys. Rev. Lett. **115**, 097802 (2015)
44. J.S.B. Tai, P.J. Ackerman, I.I. Smalyukh, Proc. Nat. Acad. Sci. **115**, 921 (2018)
45. P.M. Rupnik, D. Lisjak, M. Čopič, A. Mertelj, Liq. Cryst. **42**, 1684 (2015)
46. M. Shuai, A. Klittnick, Y. Shen, G.P. Smith, M.R. Tuchband, C. Zhu, R.G. Petschek, A. Mertelj, D. Lisjak, M. Čopič, J.E. Maclennan, M.A. Glaser, N.A. Clark, Nat. Commun. **7**, 10394 (2016)
47. P.J. Ackerman, I.I. Smalyukh, Nat. Mater. **16**, 426 (2017)
48. G. Zarubin, M. Bier, S. Dietrich, J. Chem. Phys. **149**, 054505 (2018)
49. S.D. Peroukidis, S.H. Klapp, Phys. Rev. E **92**, 010501 (2015)
50. P. Medle Rupnik, D. Lisjak, M. Čopič, A. Mertelj, Liquid Crystals **42**, 1684 (2015)
51. A.J. Bray, Adv. Phys. **51**, 481 (2002)
52. S. Puri, V. Wadhawan, *Kinetics of Phase Transitions* (CRC Press, 2009)
53. E. Ben-Jacob, I. Cohen, O. Shochet, A. Tenenbaum, A. Czirók, T. Vicsek, Phys. Rev. Lett. **75**, 2899 (1995)
54. J.K. Parrish, W.M. Hamner, *Animal Groups in Three Dimensions: How Species Aggregate* (Cambridge University Press, 1997)
55. F.J. Ndlec, T. Surrey, A.C. Maggs, S. Leibler, Nature **389**, 305 (1997)
56. D.K. Helbing, I. Farkas, T. Vicsek, Nature **407**, 487 (2000)
57. D. Helbing, I. Farkas, T. Vicsek, Phys. Rev. Lett. **84**, 1240 (2000)
58. T. Surrey, F. Nédélec, S. Leibler, E. Karsenti, Science **292**, 1167 (2001)
59. S. Hubbard, P. Babak, S. Sigurdsson, K.G. Magnússon, Ecol. Modell. **174**, 359 (2004)

References

60. C. Dombrowski, L. Cisneros, S. Chatkaew, R.E. Goldstein, J.O. Kessler, Phys. Rev. Lett. **93**, 098103 (2004)
61. W.F. Paxton, K.C. Kistler, C.C. Olmeda, A. Sen, S.K. Angelo, Y. Cao, T.E. Mallouk, P.E. Lammert, V.H. Crespi, J. Am. Chem. Soc. **126**, 13424 (2004)
62. A. Sokolov, I.S. Aranson, J.O. Kessler, R.E. Goldstein, Phys. Rev. Lett. **98**, 158102 (2007)
63. V. Schaller, C. Weber, C. Semmrich, E. Frey, A.R. Bausch, Nature **467**, 73 (2010)
64. S. Ramaswamy, Ann. Rev. Cond. Matt. Phys. **1**, 323 (2010)
65. Y. Sumino, K.H. Nagai, Y. Shitaka, D. Tanaka, K. Yoshikawa, H. Chaté, K. Oiwa, Nature **483**, 448 (2012)
66. I. Theurkauff, C. Cottin-Bizonne, J. Palacci, C. Ybert, L. Bocquet, Phys. Rev. Lett. **108**, 268303 (2012)
67. H.H. Wensink, J. Dunkel, S. Heidenreich, K. Drescher, R.E. Goldstein, H. Löwen, J.M. Yeomans, Proc. Nat. Acad. Sci. **109**, 14308 (2012)
68. J. Palacci, S. Sacanna, A.P. Steinberg, D.J. Pine, P.M. Chaikin, Science **339**, 936 (2013)
69. M.C. Marchetti, J.F. Joanny, S. Ramaswamy, T.B. Liverpool, J. Prost, M. Rao, R.A. Simha, Rev. Mod. Phys. **85**, 1143 (2013)
70. A. Sokolov, I.S. Aranson, Phys. Rev. Lett. **109**, 248109 (2012)
71. H.M. López, J. Gachelin, C. Douarche, H. Auradou, E. Clément, Phys. Rev. Lett. **115**, 028301 (2015)
72. C. Bechinger, R. Di Leonardo, H. Löwen, C. Reichhardt, G. Volpe, G. Volpe, Rev. Mod. Phys. **88**, 045006 (2016)
73. D. Takagi, A.B. Braunschweig, J. Zhang, M.J. Shelley, Phys. Rev. Lett. **110**, 038301 (2013)
74. S. Zhou, A. Sokolov, O.D. Lavrentovich, I.S. Aranson, Proc. Nat. Acad. Sci. **111**, 1265 (2014)
75. R.R. Trivedi, R. Maeda, N.L. Abbott, S.E. Spagnolie, D.B. Weibel, Soft Matter **11**, 8404 (2015)
76. C. Peng, T. Turiv, Y. Guo, Q. Wei, O.D. Lavrentovich, Science **354**, 882 (2016)
77. J.S. Lintuvuori, A. Würger, K. Stratford, Phys. Rev. Lett. **119**, 068001 (2017)
78. M.M. Genkin, A. Sokolov, O.D. Lavrentovich, I.S. Aranson, Phys. Rev. X **7**, 011029 (2017)
79. A. Sokolov, A. Mozaffari, R. Zhang, J.J. De Pablo, A. Snezhko, Phys. Rev. X **9**, 031014 (2019)
80. S. Zhou, Liq. Cryst. Today **27**, 91 (2018)
81. T. Turiv, R. Koizumi, K. Thijssen, M.M. Genkin, H. Yu, C. Peng, Q.H. Wei, J.M. Yeomans, I.S. Aranson, A. Doostmohammadi, O.D. Laverntovich, Nat. Phys. **16**, 481 (2020)
82. A. Doostmohammadi, J. Ignés-Mullol, J.M. Yeomans, F. Sagués, Nat. Comm. **9**, 3246 (2018)
83. H. Chi, M. Potomkin, L. Zhang, L. Berlyand, I.S. Aranson, Comm. Phys. **3**, 1 (2020)
84. C.W. Harvey, M. Alber, L.S. Tsimring, I.S. Aranson, New J. Phys. **15**, 035029 (2013)
85. X. Shi, H. Chaté, Y. Ma, New J. Phys. **16**, 035003 (2014)
86. A. Vats, V. Banerjee, S. Puri, Europhys. Lett. **128**, 66001 (2020)
87. A. Vats, V. Banerjee, S. Puri, Soft Matter **17**, 2659 (2021)
88. Y. Kuramoto, D. Battogtokh, Nonlinear Phenom. Complex Syst. **5**, 380 (2002)
89. D.M. Abrams, S.H. Strogatz, Phys. Rev. Lett. **93**, 174102 (2004)

Chapter 2
Theoretical Background

Abstract Coarse-grained models reveal the complex behavior of FNs and LLCs at macroscopic time and length scales. The free energy functionals for nematic, magnetic, and active systems are discussed, which serve as the foundation to understand interactions between the various components. The dynamical equations are specified for each component. Fixed-point solutions and linear stability analysis are discussed as tools to characterize equilibrium and steady-state behavior. Together with numerical techniques for solving coupled dynamical equations, the cohesive framework offers clear insights into these complex systems.

2.1 Free Energy Models

Coarse-grained phenomenological models are powerful tools for understanding systems with complex interactions. These models capture macroscopic behavior without requiring microscopic details, effectively linking theoretical predictions with experimental observations. They are particularly useful for studying phase transitions, critical phenomena, and emergent properties across various material domains. A significant advantage of coarse-grained models is their computational efficiency. Unlike detailed microscopic models that require extensive information about molecular shapes and intermolecular potentials, coarse-grained models offer a more accessible and less resource-intensive alternative. In this framework, the free energy is defined in terms of an appropriate order parameter, which captures the essential behavior of the system at a coarse-grained level. Near phase transitions, the order parameter is small, allowing the free energy to be approximated using the first few terms of a Taylor expansion. This simplification provides crucial information while maintaining analytical tractability.

This book focuses on two-component systems, defining the free energy for each component and including a coupling term to capture their interactions. In subsequent sections, we present details of the free energies and order parameters for different components of FNs and LLCs, ensuring a comprehensive understanding of the macroscopic behavior of the system. In general, coarse-grained models are advantageous over particle-based models due to their simplicity, computational efficiency, and ability to provide meaningful insights with reduced data requirements.

2.1.1 Nematic Free Energy

Landau theory defines the free energy functional based on the symmetries of order parameters without caring much about the microscopic details. As mentioned in Sect. 1.1, the nematic phase features broken rotational symmetry while maintaining translational symmetry. Neither a scalar order parameter (like density in a liquid-gas transition) nor a vector order parameter (like magnetization in a paramagnetic to ferromagnetic transition) can adequately describe the order in the nematic phase. Since the macroscopic response functions of bulk material, such as electronic polarizability, refractive index, or diamagnetic susceptibility, are tensorial and anisotropic, orientational order in the nematic phase is quantified similarly using a second-rank symmetric traceless tensor $\mathbf{Q}$. It can be defined as:

$$Q_{ij} = \mathcal{S} n_i n_j + \mathcal{T} k_i k_j - (\mathcal{S} + \mathcal{T})\frac{\delta_{ij}}{d}. \quad (2.1)$$

where the amplitude $\mathcal{S}$ is the scalar order parameter which measures the uniaxial degree of order about the leading eigenvector or the director $\mathbf{n}$ in d dimensions [1–3]. It can be approximated by the second order Legendre polynomial, $\mathcal{S} = \overline{P_2(\cos\theta)}$, where θ is the angle between the nematic molecule and the director, and the over-bar indicates an average over all the molecules. In addition, $\mathcal{T}$ is the magnitude of the biaxial order of the secondary director $\mathbf{k}$. A system with only uniaxial order has $\mathcal{T} = 0$. The tracelessness imposes the condition that material response is isotropic in the orientationally and translationally symmetric disordered fluid phase.

In terms of the $\mathbf{Q}$-tensor, the free energy for NLCs is written as:

$$\mathcal{F}_N = \int d\mathbf{r} \left[\frac{A}{2}\text{Tr}(\mathbf{Q}^2) + \frac{C}{3}\text{Tr}(\mathbf{Q}^3) + \frac{B}{4}\text{Tr}(\mathbf{Q}^2)^2 + \frac{L}{2}|\nabla\mathbf{Q}|^2 \right]. \quad (2.2)$$

The first three terms on the right-hand side are thermotropic contributions to the free energy functional. These terms dictate the state of NLCs in the equilibrium situation. For example, at high temperatures, this potential should have a minimum energy in the isotropic state, i.e., $\mathbf{Q} = 0$. On the other hand, at low temperatures, the free energy minimum corresponds to a nematic state, i.e., the eigenvalues of the $\mathbf{Q}$-tensor are finite. The gradient term $|\nabla\mathbf{Q}|^2$ captures the effect of elastic interactions and penalizes local variations in the order parameter. The parameter A has a linear dependence on the temperature T such that $A = A_0(T - T_c^N)$, where A_0 is a positive constant. The cubic term, $\text{Tr}(\mathbf{Q}^3)$, is relevant only in $d = 3$. Here, the coefficient C can take both positive and negative values. For nematic rod-like molecules $C < 0$,

2.1 Free Energy Models

whereas for disc-like molecules $C > 0$. The parameter B is a positive material-dependent constant and L is the elastic constant. The quantity B is always taken to be positive to ensure the stability and boundedness of the free energy in both the isotropic and nematic phases. These phenomenological parameters can be estimated from quantities measured experimentally. For example, A, B, C and L are related to the critical temperature, the latent heat of transition, and the magnitude of the order parameter [4]. The higher powers of $\text{Tr}(\mathbf{Q}^3)$ can be excluded for the simplest description of the uniaxial phase. The current free energy in Eq. (2.2) represents a uniaxial phase. For the biaxial phase, the higher powers of $\text{Tr}(\mathbf{Q}^3)$ are necessary [5].

In $d = 2$, the $\mathbf{Q}$-tensor has two independent components [1–3]:

$$\mathbf{Q} = \begin{pmatrix} Q_{11} & Q_{12} \\ Q_{12} & -Q_{11} \end{pmatrix}. \tag{2.3}$$

It is easy to verify that $\text{Tr}(\mathbf{Q}^2) = 2|\mathbf{Q}|^2 = 2(Q_{11}^2 + Q_{12}^2) = \mathcal{S}^2/2$ and $\text{Tr}(\mathbf{Q}^3) = 0$. In addition, the biaxial order $\mathcal{T} = 0$. The absence of the cubic term in $d = 2$ results in a *continuous* nematic-isotropic transition. The orientation of a nematogen can be fully characterized by the two independent parameters of the $\mathbf{Q}$-tensor. The isotropic phase corresponds to $\mathcal{S} = 0$, and a fully aligned nematic phase has $\mathcal{S} = 1$. A nematic defect corresponds to a region of low order or $\mathcal{S} \simeq 0$. A chosen coordinate system that diagonalizes the Q matrix at a given point in space does not automatically diagonalize the same matrix at other points in space.

In $d = 3$, the $\mathbf{Q}$ tensor has five independent parameters and, without any loss of generality, can be written as:

$$\mathbf{Q} = \begin{pmatrix} -q_1 + q_2 & q_3 & q_4 \\ q_3 & -q_1 - q_2 & q_5 \\ q_4 & q_5 & 2q_1 \end{pmatrix}. \tag{2.4}$$

It is easy to see that

$$\text{Tr}(\mathbf{Q}) = 0, \tag{2.5}$$

$$\text{Tr}(\mathbf{Q}^2) = 2q^2 = 2(3q_1^2 + q_2^2 + q_3^2 + q_4^2 + q_5^2), \tag{2.6}$$

$$\text{Tr}(\mathbf{Q}^3) = 6q_1^3 - 6q_1q_2^2 - 6q_1q_3^2 + 3q_1q_4^2 + 3q_1q_5^2 + 3q_2q_4^2 - 3q_2q_5^2$$
$$+ 6q_3q_4q_5, \tag{2.7}$$

$$\text{Tr}(\mathbf{Q}^2)^2 = 4(3q_1^2 + q_2^2 + q_3^2 + q_4^2 + q_5^2)^2. \tag{2.8}$$

The presence of a cubic term leads to a first-order phase transition in the NLCs, and has important consequences on the characteristics of the topological defects. A frame of reference that diagonalizes $\mathbf{Q}$ provides the three eigenvalues ($\lambda_3 > \lambda_2 > \lambda_1$), and the corresponding eigenvectors $\mathbf{n}$, $\mathbf{k}$, $\mathbf{l}$. The largest eigenvalue $\lambda_3 = \mathcal{S}$, and the corresponding eigenvector is the primary direction of order $\mathbf{n}$ [6, 7]. If $\lambda_1 = \lambda_2$, the system is uniaxial. The degree of biaxial order for the secondary director **k** is given by $\mathcal{J} = (\lambda_2 - \lambda_1)/\lambda_3$ [7–9]. It can also be defined as $\mathcal{B}^2 = (1 - 6\mathrm{Tr}(\mathbf{Q}^3)^2/(\mathrm{Tr}(\mathbf{Q}^2)^3))$ [10, 11]. A value of $\mathcal{B}^2 = 0$ indicates uniaxiality while $\mathcal{B}^2 = 1$ corresponds to a state with maximum biaxiality.

2.1.2 Ginzburg–Landau Free Energy for Magnetic System

The free energy for the magnetic components is associated with the average magnetic moment of MNPs **M**, and can be written as

$$\mathcal{F}_M(\mathbf{M}) = \int d\mathbf{r} \left[\frac{\alpha^M}{2} |\mathbf{M}|^2 + \frac{\beta^M}{4} |\mathbf{M}|^4 + \frac{\kappa^M}{2} |\nabla \mathbf{M}|^2 \right]. \tag{2.9}$$

This is the standard Ginzburg–Landau (GL) free energy, where α^M, β^M and κ^M are Landau coefficients describing the ferromagnetic transition as in Ref. [12]. The superscript 'M' here denotes the magnetic system. The parameter $\alpha^M = \alpha_0^M(T - T_c^M)$, where α_0^M is the material-dependent constant and T_c^M is the critical temperature. The coefficient β^M is a positive constant, and κ^M is the elastic constant related to magnetic stiffness. The elastic term smoothens the short-range variations or fluctuations in **M** and ensures that there are no spatial inhomogeneities in the order parameter. When α^M is positive, the free energy functional exhibits a minimum at $|\mathbf{M}| = 0$, corresponding to an isotropic state. In contrast, for $\alpha^M < 0$ the free energy is minimized to a state with magnetic order corresponding to $|\mathbf{M}| = \sqrt{\alpha^M/\beta^M}$ (see Fig. 2.1). This solution characterizes a ferromagnetic state with rotational symmetry. A magnetic

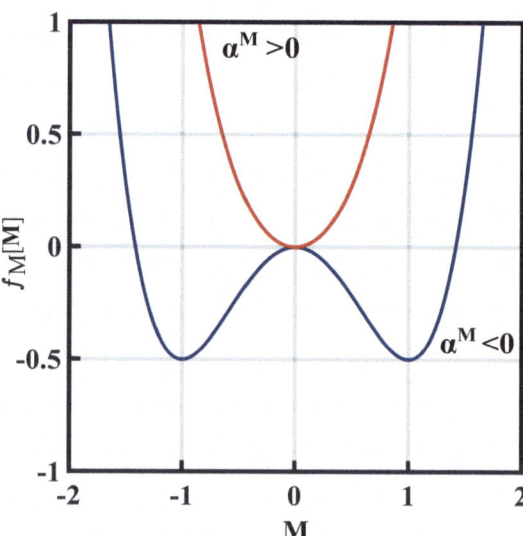

Fig. 2.1 Plot depicting the free energy minima for the order parameter **M**

2.1 Free Energy Models

system quenched below its critical temperature undergoes spontaneous symmetry breaking and settles into a uniform state aligned along one of the directions among the rotationally symmetric degenerate states. The $\mathbf{M}^4$ and higher-order terms ensure the stability of the free energy functional and prevent it from becoming unbounded. These Landau parameters can be evaluated in experiments from susceptibility and magnetization measurements [13]. For the $d = 2$ system, it is assumed that the MNPs primarily have an in-plane alignment of magnetic moments. Therefore, $\mathbf{M}$ is a vector with two (three) components in $d = 2$ (3). Furthermore, $\mathbf{M}$ can have a variable magnitude. $\mathbf{M} = 0$ state corresponds to a defect and $\mathbf{M} = 1$ is a fully aligned state in the system.

2.1.3 Free Energy for Active Particles

In active systems, the free energy serves a different role than in traditional equilibrium contexts. In equilibrium, free energy represents the potential for the system to perform work while moving toward or maintaining a minimum energy configuration. However, active systems inherently remain out of equilibrium because they continuously consume and dissipate energy. As a result, their behavior cannot be fully captured by a free energy functional alone. For active systems, we can still define a *downhill dynamics* using a pseudo free energy functional, which captures the system's tendency to minimize certain energy contributions. To address the ongoing energy input and dissipation, additional nonequilibrium terms are added to the dynamical equations. These terms represent the active forces driving the system away from equilibrium, reflecting the influence of continuous energy consumption on the evolution of the system. Thus, the combined formalism of the free-energy and nonequilibrium contributions enables a coherent approach to studying the complex dynamics of active systems.

The order parameters for the equations of motion are (i) the local density of the active particles $\rho(\mathbf{r}, t)$ and (ii) the local polarization $\mathbf{P}(\mathbf{r}, t)$ that describes their average orientation [14–18]. The free energy functional is given by [16, 18]:

$$\mathcal{F}_a[\rho, \mathbf{P}] = \int d\mathbf{r} \left[\frac{\alpha(\rho)}{2}|\mathbf{P}|^2 + \frac{\beta}{4}|\mathbf{P}|^4 + \frac{\kappa}{2}|\nabla \mathbf{P}|^2 + \frac{w}{2}|\mathbf{P}|^2 \nabla \cdot \mathbf{P} \right.$$
$$\left. - \frac{v_1}{2}(\nabla \cdot \mathbf{P})\frac{\delta\rho}{\rho_0} + \frac{D_\rho}{2}(\delta\rho)^2 \right], \quad (2.10)$$

Here, the parameters α, β, κ, w, v_1, and D_ρ are material-dependent and connect to the microscopic properties of active particles, providing a macroscopic description of the system [19, 20]. Here, the parameter $\alpha(\rho) = \alpha_0 (1 - \rho/\rho_c)$ is density dependent, where ρ_c represents the critical density for observing the orientational order in the active system. The gradient term, $|\nabla \mathbf{P}|^2$, captures the energy cost associated with spatial variations in the order parameter, modeling resistance to deformation within the orientation field. The following terms introduce the effects of orientational and density fields on spontaneous splay ($\nabla \cdot \mathbf{P}$). These terms, linked to $|\mathbf{P}|^2$ and density, represent local alignment forces generated by variations in density and the magnitude of the order parameter. Finally, the term $(\delta\rho)^2$, where $\delta\rho = \rho - \rho_0$, penalizes deviations from the mean density ρ_0. This term helps stabilize the density around ρ_0, maintaining a uniform background density while allowing for local fluctuations. A more comprehensive discussion of the physical interpretation and relevance of each term can be found in [16, 18].

2.2 Dynamical Equations

The dynamical equations capture the temporal evolution of the system as it approaches its free energy minimum. These equations inherently define the system's relaxation timescales, balancing the influence of competing interactions on the evolution. In the subsections that follow, the equations of motion are presented for each of the uncoupled components, namely the NLCs, MNPs, and AM.

2.2.1 Time-Dependent Ginzburg–Landau Equations

The framework introduced in Sects. 2.1.1 and 2.1.2 allow the nematic and magnetic order parameters, represented as $\mathbf{Q}$ and $\mathbf{M}$, to vary in magnitude. The behavior of both systems is described by a non-conserved dynamics. On the microscopic scale, Ising-type models with Glauber spin-flip dynamics are commonly used to effectively capture the key aspects of phase transitions and self-assembly. However, direct particle simulations of such models are computationally intensive when scaled up to experimentally relevant system sizes. To overcome this, a coarse-grained approach is employed using the time-dependent Ginzburg–Landau equations [12, 21]. These equations describe how the system dissipatively relaxes toward its free energy minimum. Although the TDGL framework is rigorously valid only near phase transitions, where the order parameter remains small, studies show that the resulting equations can effectively capture the essential physical processes even for deep quenches ($T \ll T_c$). For a general order parameter ψ, the TDGL equation is expressed as [12, 21]:

$$\frac{\partial \psi(\mathbf{r}, t)}{\partial t} = -\Gamma_\psi \frac{\delta \mathcal{F}_\psi[\psi]}{\delta \psi}, \quad (2.11)$$

where the terms on the right are functional derivatives of the free energy [12]. The order parameter ψ can take various forms, including scalar, vector, or tensor, depending on the specific system studied. In the case of nematic and magnetic components, it corresponds to $\mathbf{Q}$ and $\mathbf{M}$, respectively. The relaxation time scale of the system is set

2.2 Dynamical Equations

by the damping factor Γ_ψ. Note that there is no noise term in Eq. (2.11) indicating an effective temperature $T = 0$. Using renormalization group theory, it can be proved that temperatures above T_c flow to infinity, while those below T_c flow to zero [21–23]. As a result, the final temperature becomes an irrelevant variable for quenches in the ordered phase.

2.2.2 Toner–Tu Equations for Active Particles

The dynamics of active particles has proven to be highly intriguing from a fundamental physics perspective because of their unique properties that differ markedly from equilibrium systems. These systems offer a distinctive platform to investigate various hypotheses related to universality and scaling behavior. One of the pioneering efforts to model such an active system was introduced by Vicsek et al., who employed direct particle simulations to explore the collective dynamics of active particles [24]. Their model is notable for its ability to effectively capture and explain key features of active systems, such as phase transitions and self-organization. It provides valuable insights into how individual particles' interactions lead to emergent collective behavior, such as flocking or pattern formation. However, the primary limitation of Vicsek's model is its computational cost. The need for detailed simulations of many interacting particles means that the model is generally applicable only to relatively small systems.

A macroscopic description of such systems demands a continuum approach based on the mesoscopic order parameter. This formalism is provided by the elegant hydrodynamic theory of Toner and Tu (TT) [14, 15]. Although the original model is formulated phenomenologically using symmetry considerations, it is instructive to rewrite the equations of motion in terms of a free energy functional $F_a[\rho, \mathbf{P}]$ [16, 18]:

$$\frac{\partial \rho}{\partial t} = -v_0 \nabla \cdot (\mathbf{P}\rho) - \nabla \cdot \left(-\Gamma_\rho \nabla \frac{\delta F_a}{\delta \rho}\right), \quad (2.12)$$

$$\frac{\partial \mathbf{P}}{\partial t} = \lambda_1 (\mathbf{P} \cdot \nabla) \mathbf{P} - \Gamma_P \frac{\delta F_a}{\delta \mathbf{P}}. \quad (2.13)$$

Here, v_0 is the speed of the active particles, and Γ_ρ and Γ_P set the relaxation time scales for the density and polarization fields. Eq. (2.12) is the continuity equation for the density field and represents the conserved nature of the fields. The first term here quantifies the change in the density due to the polarization field and couples the polarization vector to the density. Also note that the coefficient of this term is the speed v_0. Hence, the polarization enters the density equation as the velocity field. The second term of the equation yields a familiar diffusion current. In the TT model, the field $\mathbf{P}$ acts both as a current and as an orientational order parameter. Hence, it evolves in time [Eq. (2.13)] via both advection and flow alignment. In addition, the parameter λ_1, which controls the strength of the advective term, has units of speed. This term is similar to the advective non-linearity found in the Navier-Stokes equation. However, in the nonequilibrium flocking model, there is no conservation of momentum since the particles move relative to a substrate. Furthermore, the model is not restricted by Galilean invariance, which would otherwise require $\lambda_1 = v_0$. Therefore, λ_1 serves as a non-universal phenomenological parameter influenced by the specific microscopic properties of the system. This allows for the convection of polarization and density fluctuations at different speeds. The last term in Eq. (2.13) takes the system to its free energy minimum.

2.3 Numerical Techniques Used to Solve Dynamical Equations

The free energy models (Sect. 2.1) and corresponding dynamical equations (Sect. 2.2) are used to determine morphologies for these coupled systems. These equations are solved numerically with the Euler discretization method [25], where the spatial and temporal scales are set by the mesh sizes Δx and Δt. As long as certain conditions are met, such as (a) adhering to the stability criterion and (b) ensuring that Δx adequately captures the defect region, these mesh choices do not alter the nature of the solutions [26–29]. In this approach, first and second-order derivatives of a function $f(x, t)$ are discretized using Euler formulae as [30]:

$$\frac{\partial f}{\partial t} = \frac{f(t + \Delta t) - f(t)}{\Delta t}, \tag{2.14}$$

$$\frac{\partial f}{\partial x} = \frac{f(x + \Delta x) - f(x - \Delta x)}{2\Delta x}, \tag{2.15}$$

$$\frac{\partial^2 f}{\partial^2 x} = \frac{f(x + \Delta x) + f(x - \Delta x) - 2f(x)}{(\Delta x)^2}. \tag{2.16}$$

Although various discretization schemes and sophisticated techniques can provide more accurate approximations of partial derivatives, the current method effectively captures the essential physical behaviors. To derive the stability condition for the TDGL equation, consider the dimensionless version of Eq. (2.11) with the free energy in Eq. (2.9). The dimensionless form is obtained via a suitable rescaling of space, time and order parameter. We consider the following finite-difference scheme for this equation:

$$\psi(\mathbf{r}, t + \Delta t) = \psi(\mathbf{r}, t) + \Delta t \left[\psi(\mathbf{r}, t) - \psi(\mathbf{r}, t)^3\right] + \frac{\Delta t}{(\Delta x)^2} \Delta_D \psi(\mathbf{r}, t), \tag{2.17}$$

where Δ_D represents the discrete Laplacian operator. The finite-difference equation (2.17) can be solved numerically for any given initial condition by setting $\psi(\mathbf{r}, t)$ at each point in the grid. To ensure that this finite-difference scheme accurately reflects the physics of the original partial differential equation, it is crucial that the behavior of the domains within the bulk remains stable, converging toward equilibrium values of $+1$ or -1. Thus, any small fluctuations around these equilibrium values must not grow over time.

The stability is assessed by substituting $\psi(\mathbf{r}, t) = 1 + \delta\psi(\mathbf{r}, t)$ into Eq. (2.17) and verifying that, under a linear approximation, these fluctuations diminish as the system evolves. The resulting equations are:

$$\delta\psi(\mathbf{r}, t + \Delta t) = (1 - 2\Delta t)\delta\psi(\mathbf{r}, t) + \{\Delta t/(\Delta x)^2\}\Delta_D \delta\psi(\mathbf{r}, t), \quad (2.18)$$

Multiplying both sides by $e^{i\mathbf{k}\cdot\mathbf{r}}$ and summing over $\mathbf{r}$ yields:

$$\delta\psi(\mathbf{k}, t + \Delta t) = \left(1 - 2\Delta t - 2\{\Delta t/(\Delta x)^2\}[2 - \cos(k_x \Delta x) - \cos(k_y \Delta x)]\right)\delta\psi(\mathbf{k}, t),$$
$$= \lambda(\mathbf{k})\delta\psi(\mathbf{k}, t). \quad (2.19)$$

The stability of the solution requires $|\lambda(\mathbf{k})| < 1$ which suggests unphysical divergences (sub-harmonic bifurcation) for $\lambda(\mathbf{k}) < -1$. This yields the stability condition as [30]:

$$\Delta t < \frac{\Delta x^2}{\Delta x^2 + 4}. \quad (2.20)$$

The above criterion is for the two-dimensional system, and it is easy to verify that, for a d-dimensional system, 4 gets replaced by $2d$. The same condition applies to the vector TDGL equation and is used to determine the mesh sizes in the simulations.

References

1. C. Luo, A. Majumdar, R. Erban, Phys. Rev. E **85**, 061702 (2012)
2. K. Bisht, V. Banerjee, P. Milewski, A. Majumdar, Phys. Rev. E **100**, 012703 (2019)
3. K. Bisht, Y. Wang, V. Banerjee, A. Majumdar, Phys. Rev. E **101**, 022706 (2020)
4. E. Priestly, *Introduction to Liquid Crystals* (Springer Science, 2012)
5. A. Majumdar, Eur. J. Appl. Math. **21**, 181 (2010)
6. N.J. Mottram, J.P. Newton (2014). arXiv:1409.3542
7. A. Bhattacharjee, Inhomogeneous phenomena in nematic liquid crystals. Ph.D thesis (2010)
8. C. Tschierske, D.J. Photinos, J. Mater. Chem. **20**, 4263 (2010)
9. G.R. Luckhurst, T.J. Sluckin, *Biaxial Nematic Liquid Crystals: Theory, Simulation and Experiment* (Wiley, 2015)
10. P. Kaiser, W. Wiese, S. Hess, J. Non-Equilib, Thermodyn. **17**, 153 (1992)
11. S. Kralj, R. Rosso, E.G. Virga, Phys. Rev. E **81**, 021702 (2010)
12. S. Puri, V. Wadhawan, *Kinetics of Phase Transitions* (CRC Press, 2009)
13. P.C. Hohenberg, A.P. Krekhov, Phys. Rep. **572**, 1 (2015)
14. J. Toner, Y. Tu, Phys. Rev. Lett. **75**, 4326 (1995)
15. J. Toner, Y. Tu, Phys. Rev. E **58**, 4828 (1998)
16. S. Ramaswamy, Ann. Rev. Cond. Matt. Phys. **1**, 323 (2010)
17. S. Mishra, A. Baskaran, M.C. Marchetti, Phys. Rev. E **81**, 061916 (2010)
18. M.C. Marchetti, J.F. Joanny, S. Ramaswamy, T.B. Liverpool, J. Prost, M. Rao, R.A. Simha, Rev. Mod. Phys. **85**, 1143 (2013)

19. E. Bertin, M. Droz, G. Grégoire, Phys. Rev. E **74**, 022101 (2006)
20. E. Bertin, M. Droz, G. Grégoire, J. Phys. A **42**, 445001 (2009)
21. A.J. Bray, Adv. Phys. **51**, 481 (2002)
22. A.J. Bray, Phys. Rev. Lett. **62**, 2841 (1989)
23. A.J. Bray, Phys. Rev. B **41**, 6724 (1990)
24. T. Vicsek, A. Czirók, E. Ben-Jacob, I. Cohen, O. Shochet, Phys. Rev. Lett. **75**, 1226 (1995)
25. E.W. Cheney, D.R. Kincaid, *Numerical Mathematics and Computing* (Cengage Learning, 2012)
26. Y. Oono, S. Puri, Phys. Rev. Lett. **58**, 836 (1987)
27. Y. Oono, S. Puri, Phys. Rev. A **38**, 434 (1988)
28. S. Puri, Y. Oono, Phys. Rev. A **38**, 1542 (1988)
29. K.R. Elder, T.M. Rogers, R.C. Desai, Phys. Rev. B **38**, 4725 (1988)
30. S. Puri, Kinetics of phase transitions: Numerical techniques and simulations, in *Computational Statistical Physics*, ed. by S.B. Santra, P. Ray (Springer, 2011), p. 123

Chapter 3
Phase Ordering Dynamics in Ferronematics

Abstract The nonequilibrium phenomenon of *phase ordering kinetics* (or *domain growth* or *coarsening*) after a temperature quench is studied in FNs. The focus here is to investigate the role of coupling in pattern formation. The modeling is based on *time-dependent Ginzburg–Landau* equations for the coupled order parameters. Quenches from a high-temperature disordered phase are considered. Subsequently, the system coarsens by collision and annihilation of topological defects. A notable phenomenon, termed *slaved coarsening*, occurs when an intrinsically disordered component is driven to coarsen by an ordered one. Extensive numerical and analytical studies of morphologies, defect dynamics, and growth laws reveal unique features arising purely from magneto-nematic coupling.

3.1 Introduction

A well-studied nonequilibrium process is the *kinetics of phase transitions*, which occurs when a system is quenched from a disordered state ($T > T_c$) to an ordered state ($T < T_c$). The system does not achieve order immediately. Instead, regions of degenerate ground states emerge, separated by interfaces or defects [1, 2]. Over time, these domains *coarsen* as defects are eliminated, leading to the development of a characteristic length scale $L(t)$ that increases over time. The laws governing domain growth provide crucial insights into the system's ordering, with late-stage growth driven by defect dynamics. Figure 3.1 schematically illustrates this coarsening process in a ferromagnetic system.

The investigations of phase ordering in $d = 2$ thermotropic NLCs commenced in the 1990's and continue to draw considerable attention [2–10]. The primary reason for this ongoing interest is that the order parameter **Q** is not a simple scalar or vector, but a traceless symmetric tensor that provides insight into the directional properties of NLCs [11]. Moreover, NLCs are experimentally realized systems that exhibit continuous symmetry, supporting stable topological defects with integer or half integer charges that are eliminated in the coarsening process [4, 5]. The dynamics of coarsening in the late stages, governed by these defects, is different from those of models with $O(n)$ symmetry. The latter are characterized by vector order parameters

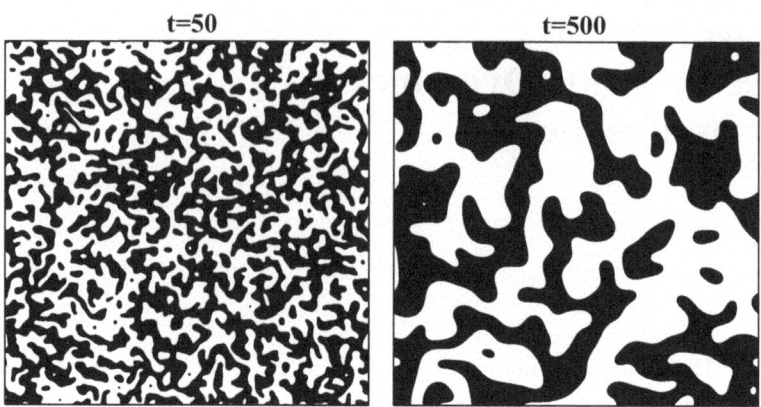

Fig. 3.1 Coarsening process for a ferromagnetic system quenched below its critical temperature

with n components but do not have the inversion symmetry of the NLC director field **n** [2, 4]. (Illustrations of defects in NLCs and analogous magnetic systems are available in Refs. [12–15].)

This chapter presents benchmark studies of coarsening in $d = 2, \ 3$ FNs [16, 17]. The approach here is based on the GL free energies and the corresponding relaxation dynamics described by the TDGL equations. Two-dimensional geometries have been experimentally accessible in pure NLCs in shallow wells where the molecules are primarily confined to a plane. This is accomplished by applying planar boundary conditions to the top and bottom surfaces, which constrain the molecules to remain within a single plane. In addition, the methodology has also been extended for the $d = 3$ system with intricate defect structures, including hedgehogs and strings [12, 18]. The complexity of the $d = 3$ system arises from the need to solve five coupled partial differential equations. This increased computational challenge may explain why coarsening in $d = 3$ systems has not been studied as extensively as in lower dimensions [4, 19, 20]. When MNPs are introduced, the complexity of the system increases further, requiring the solution of eight coupled partial differential equations, and this makes the analysis more challenging. Here, the study provides a detailed discussion of numerical techniques, together with comprehensive analytical results for $d = 2$ and extensive numerical results for both $d = 2$ and $d = 3$.

The chapter is structured as follows. Section 3.2 introduces the GL free energy models for FNs and derives the associated TDGL equations. It includes an analysis of fixed points and their linear stability in $d = 2$. In addition, the section covers the tools necessary for characterizing morphologies and deriving growth laws. Section 3.3 presents detailed numerical results for $d = 2$ and $d = 3$ FNs. The chapter concludes with a summary and discussion of the findings in Sect. 3.4.

3.2 Theoretical Framework

3.2.1 Ginzburg–Landau Free Energy

The free energy for FNs has contributions from the nematic and magnetic components along with a suitable coupling term to capture their interactions. The GL free energy for the FNs has the form [12, 21–24]:

$$G(\mathbf{Q}, \mathbf{M}) = \int d\mathbf{r} \left[\frac{A}{2} \text{Tr}(\mathbf{Q}^2) + \frac{C}{3} \text{Tr}(\mathbf{Q}^3) + \frac{B}{4} \text{Tr}(\mathbf{Q}^2)^2 + \frac{L}{2} |\nabla \mathbf{Q}|^2 \right.$$
$$\left. + \frac{\alpha^M}{2} |\mathbf{M}|^2 + \frac{\beta^M}{4} |\mathbf{M}|^4 + \frac{\kappa^M}{2} |\nabla \mathbf{M}|^2 - \frac{\gamma \mu_0}{2} \sum_{i,j=1}^{3} Q_{ij} M_i M_j \right]. \quad (3.1)$$

A similar form of free energy has also been proposed by Mertelj et al. in terms of $\mathbf{n}$ and $\mathbf{M}$ [22]. The first four terms in Eq. (3.1) represent the LdG free energy for the nematic component and the next three terms correspond to the GL free energy for the magnetic component. The interaction between the nematic and magnetic components is described by the dyadic product of $\mathbf{Q}$ and $\mathbf{M}$. Experimentally, the two components tend to align, and this coupling represents the leading-order term that respects both the up-down symmetry and the rotational invariance of the system. Furthermore, in terms of the director, the coupling term effectively reduces to $(\mathbf{n} \cdot \mathbf{M})^2$ as proposed in Ref. [22]. The phenomenological parameter γ has been experimentally estimated using reversal fields of hysteresis loops [22, 25]. A magnetic field term is excluded from the formulation to avoid introducing directional bias, which could overshadow the effects of the magneto-nematic coupling. Furthermore, the stray field energy is not included in Eq. (3.1) since it is 1–2 orders of magnitude smaller than the contribution of the nematic elastic energy [25, 26]. Therefore, the present framework applies to the FNs obtained by Method 2 discussed in Sect. 1.1.

3.2.2 Time-Dependent Ginzburg–Landau Equations

The dynamics of FNs is modeled using TDGL equations, as explained in Sect. 2.2. For the free energy in Eq. (3.1), we obtain the corresponding TDGL equations. In $d = 2$, these can be expressed as:

$$\frac{1}{\Gamma_Q}\frac{\partial Q_{11}}{\partial t} = \pm 2|A|Q_{11} - 4B|\mathbf{Q}|^2 Q_{11} + 2L\nabla^2 Q_{11} + \frac{\gamma\mu_0}{2}\left(M_1^2 - M_2^2\right), \quad (3.2)$$

$$\frac{1}{\Gamma_Q}\frac{\partial Q_{12}}{\partial t} = \pm 2|A|Q_{12} - 4B|\mathbf{Q}|^2 Q_{12} + 2L\nabla^2 Q_{12} + \gamma\mu_0\left(m_1 m_2\right), \quad (3.3)$$

$$\frac{1}{\Gamma_M}\frac{\partial M_1}{\partial t} = \pm|\alpha^M|M_1 - \beta^M|\mathbf{M}|^2 M_1 + \kappa^M \nabla^2 M_1 + \gamma\mu_0\left(Q_{11}M_1 + Q_{12}M_2\right), \quad (3.4)$$

$$\frac{1}{\Gamma_M}\frac{\partial M_2}{\partial t} = \pm|\alpha^M|M_2 - \beta^M|\mathbf{M}|^2 m_2 + \kappa^M \nabla^2 M_2 + \gamma\mu_0\left(Q_{12}M_1 - Q_{11}M_2\right). \quad (3.5)$$

The symbols $+(-)$ indicate whether the quench temperature is below or above the critical temperature T_C, respectively. Note that the TDGL equations used here are deterministic, excluding thermal fluctuations or noise. This is because noise becomes asymptotically irrelevant in domain growth problems [27], except when quenched disorder is present [28, 29].

The number of independent parameters can be reduced by appropriate rescaling of the order parameters, space, and time. Introducing the rescaled variables $\mathbf{Q} = a\mathbf{Q}'$, $\mathbf{M} = b\mathbf{M}'$, $\mathbf{r} = \xi\mathbf{r}'$, and $t = \tau t'$ leads to a dimensionless form of the TDGL equations. The appropriate scale factors are $a = \sqrt{|A|/2B}$, $b = \sqrt{|\alpha^M|/\beta^M}$, $\xi = \sqrt{\kappa^M/|\alpha^M|}$, $\tau = 1/(|\alpha^M|\Gamma_M)$. Substituting these into Eqs. (3.2)–(3.5) and omitting the primes, the resulting TDGL equations are as follows [16, 17]:

$$\frac{1}{\Gamma}\frac{\partial Q_{11}}{\partial t} = \pm Q_{11} - |\mathbf{Q}|^2 Q_{11} + l\nabla^2 Q_{11} + c_1\left(M_1^2 - M_2^2\right), \quad (3.6)$$

$$\frac{1}{\Gamma}\frac{\partial Q_{12}}{\partial t} = \pm Q_{12} - |\mathbf{Q}|^2 Q_{12} + l\nabla^2 Q_{12} + 2c_1\left(M_1 M_2\right), \quad (3.7)$$

$$\frac{\partial M_1}{\partial t} = \pm M_1 - |\mathbf{M}|^2 M_1 + \nabla^2 M_1 + c_2\left(Q_{11}M_1 + Q_{12}M_2\right), \quad (3.8)$$

$$\frac{\partial M_2}{\partial t} = \pm M_2 - |\mathbf{M}|^2 M_2 + \nabla^2 M_2 + c_2\left(Q_{12}M_1 - Q_{11}M_2\right). \quad (3.9)$$

The dimensionless parameters in Eqs. (3.6)–(3.9) are

$$c_1 = \frac{\gamma\mu_0|\alpha^M|}{4|A|\beta^M}\sqrt{\frac{2B}{|A|}}, \quad c_2 = \frac{\gamma\mu_0}{|\alpha^M|}\sqrt{\frac{|A|}{2B}}, \quad l = \frac{|\alpha^M|L}{2|A|\kappa^M}, \quad \Gamma = \frac{2|A|\Gamma_Q}{|\alpha^M|\Gamma_M}. \quad (3.10)$$

The rescaled coupling constants c_1 and c_2 quantify the interactions between the nematic and magnetic components: c_1 reflects the influence of $\mathbf{M}$ on $\mathbf{Q}$, while c_2 measures the impact of $\mathbf{Q}$ on $\mathbf{M}$. Both constants originate from the same coupling in Eq. (3.1), but during the dimensionless re-scaling, they are combined with factors that define the dimensional scales for the order parameters $\mathbf{Q}$ and $\mathbf{M}$ [see Eq. (3.10)]. The parameter l determines the scale for the relative diffusion of these components, while Γ is the relative damping coefficient. Importantly, l modifies only the non-universal prefactors of the growth laws without affecting the universal growth exponents. These exponents depend on several factors: (a) system dynamics (whether conserved or

3.2 Theoretical Framework

non-conserved), (b) the nature of defects driving coarsening, and (c) the influence of hydrodynamic effects [1, 2]. To simplify the calculations, l and Γ are set to 1. However, the relaxation time scales of $\mathbf{Q}$ and $\mathbf{M}$ can still vary significantly as these are scaled parameters. The signs of the first terms on the right depend on whether the order parameters ($\mathbf{Q}$ or $\mathbf{M}$) are above $(-)$ or below $(+)$ their critical temperatures. Typically, a nematic system is characterized by the ratios of its intrinsic length or time scales. The relevant length scale for a pure nematic system is the correlation length of the defect core, given as: $\zeta \sim \sqrt{L/|A|}$. The dimensionless scaled units can be constructed using the ratios of nematic parameters (A, B, C) or elastic constants.

Note that the $\mathbf{Q}$-tensor in $d = 3$ has five independent components, and the cubic term in the free energy is non-zero (see Sect. 2.2). The scaled equations are derived by applying a similar rescaling procedure used for $d = 2$. The suitable scale factors for this case are: $a = \sqrt{|A|/2B}$, $b = \sqrt{|\alpha^M|/\beta^M}$, $\xi = \sqrt{\kappa^M/|\alpha^M|}$, $\tau = 1/(|\alpha^M|\Gamma_M)$. Dropping the primes, the TDGL equations in $d = 3$ become

$$\frac{1}{\Gamma}\frac{\partial q_1}{\partial t} = \pm 3q_1 - q^2 3q_1 - \bar{C}(6q_1^2 - 2q_2^2 - 2q_3^2 + q_4^2 + q_5^2) + l\nabla^2 q_1$$
$$+ c_1(-M_1^2 - M_2^2 + 2M_3^2), \qquad (3.11)$$

$$\frac{1}{\Gamma}\frac{\partial q_2}{\partial t} = \pm q_2 - q^2 q_2 - \bar{C}(4q_1 q_2 + q_4^2 - q_5^2) + l\nabla^2 q_2 + c_1(M_1^2 - M_2^2), \qquad (3.12)$$

$$\frac{1}{\Gamma}\frac{\partial q_3}{\partial t} = \pm q_3 - q^2 q_3 - \bar{C}(-4q_1 q_3 + 2q_4 q_5) + l\nabla^2 q_3 + 2c_1 M_1 M_2, \qquad (3.13)$$

$$\frac{1}{\Gamma}\frac{\partial q_4}{\partial t} = \pm q_4 - q^2 q_4 - \bar{C}(2q_1 q_4 + 2q_2 q_4 + 2q_3 q_5) + l\nabla^2 q_4 + 2c_1 M_1 M_3, \qquad (3.14)$$

$$\frac{1}{\Gamma}\frac{\partial q_5}{\partial t} = \pm q_5 - q^2 q_5 - \bar{C}(2q_1 q_5 - 2q_2 q_5 + 2q_3 q_4) + l\nabla^2 q_5 + 2c_1 M_2 M_3, \qquad (3.15)$$

$$\frac{\partial M_1}{\partial t} = \pm M_1 - |\mathbf{M}|^2 M_1 + \nabla^2 M_1 + c_2[(q_2 - q_1)M_1 + q_3 M_2 + q_4 M_3], \qquad (3.16)$$

$$\frac{\partial M_2}{\partial t} = \pm M_2 - |\mathbf{M}|^2 M_2 + \nabla^2 M_2 + c_2[-(q_1 + q_2)M_2 + q_3 M_1 + q_5 M_3], \qquad (3.17)$$

$$\frac{\partial M_3}{\partial t} = \pm M_3 - |\mathbf{M}|^2 M_3 + \nabla^2 M_3 + c_2[2q_1 M_3 + q_4 M_1 + q_5 M_2]. \qquad (3.18)$$

Here, the dimensionless parameters are:

$$\Gamma = \frac{2|A|\Gamma_Q}{|\alpha^M|\Gamma_m}, \quad \bar{C} = \frac{C}{2\sqrt{2|A|B}}, \quad l = \frac{L|\alpha^M|}{2\kappa^M|A|},$$
$$c_1 = \frac{\gamma\mu_0|\alpha^M|}{4\beta^M|A|}\sqrt{\frac{2B}{|A|}}, \quad c_2 = \frac{\gamma\mu_0}{2|\alpha^M|}\sqrt{\frac{|A|}{2B}}. \qquad (3.19)$$

There are three cases of potential interest in this problem: (1) $T_c^N < T < T_c^M$, (2) $T_c^M < T < T_c^N$, and (3) $T < \min\{T_c^N, T_c^M\}$. Furthermore, each of these has subcases corresponding to different coupling limits as specified in Table 3.1. Some comments on the limiting cases are as follows. Asymmetric coupling is common

Table 3.1 Coarsening studies which have been undertaken. Reprinted with permission from [17]. © 2021, Royal Society of Chemistry. All rights reserved

Quench temperature	Coupling constants
(1) $T_c^N < T < T_c^M$	(i) $c_1 \neq 0, c_2 = 0$
(2) $T_c^M < T < T_c^N$	(ii) $c_1 = 0, c_2 \neq 0$
(3) $T < \min\{T_c^N, T_c^M\}$	(iii) $c_1 = c_2 = c$

in experimental systems, where the order parameters can differ significantly in magnitude, such as large magnetic particles in a bath of smaller liquid crystal molecules. Although it is unlikely that specific cases like $c_2 = 0$, $c_1 = 0$, or $c_1 = c_2 = c$ would occur exactly in experiments, scenarios where $c_1 \gg c_2$ or $c_2 \gg c_1$ are more plausible. These correspond to the cases (i) and (ii) discussed above. Despite their simplicity, such limiting cases offer valuable insight into theoretical models.

3.2.3 Fixed-Point Solutions for $d = 2$ FNs

Equations (3.6)–(3.9) govern the relaxation dynamics of **Q** and **M** in $d = 2$. Although the primary focus is on coarsening, it is instructive to examine the long-term asymptotic behavior or stationary solutions (**Q***, **M***) of Eqs. (3.6)–(3.9). As there is no spatial or temporal variation in this limit, they can be obtained by setting $\partial \psi / \partial t$ and $\nabla^2 \psi$ to 0 in Eqs. (3.6)–(3.9) which yields:

$$\pm Q^*_{11} - |\mathbf{Q}^*|^2 Q^*_{11} + c_1(M^*_1{}^2 - M^*_2{}^2) = 0, \qquad (3.20)$$

$$\pm Q^*_{12} - |\mathbf{Q}^*|^2 Q^*_{12} + 2c_1(M^*_1 M^*_2) = 0, \qquad (3.21)$$

$$\pm M^*_1 - |\mathbf{M}^*|^2 M^*_1 + c_2(Q^*_{11} M^*_1 + Q^*_{12} M^*_2) = 0, \qquad (3.22)$$

$$\pm M^*_2 - |\mathbf{M}^*|^2 M^*_2 + c_2(Q^*_{12} M^*_1 - Q^*_{11} M^*_2) = 0. \qquad (3.23)$$

A trivial solution is $Q^*_{11} = Q^*_{12} = M^*_1 = M^*_2 = 0$, which represents a disordered state at high temperatures. The explicit non-trivial solutions for cases 1 to 3 in Table 3.1 are detailed in Tables A.1–A.3. Fixed points can also be obtained analytically for any value of c_1 and c_2, but the expressions are quite cumbersome and are not presented here.

To determine stability, the evolution of small perturbations around the stationary solutions (**Q*** + Δ**Q**, **M*** + Δ**M**) are studied using the TDGL equations (3.6)–(3.9). It is convenient to work with the Fourier transforms [Δ**Q**(**k**, t), Δ**M**(**k**, t)], which in the linear approximation results in the corresponding equations:

3.2 Theoretical Framework

$$\frac{\partial \Delta Q_{11}}{\partial t} = 2c_1 M_1^* \Delta M_1 - 2c_1 M_2^* \Delta M_2 + \left(\pm 1 - 3Q_{11}^{*2} - Q_{12}^{*2} - \mathbf{k}^2\right) \Delta Q_{11}$$
$$- 2Q_{11}^* Q_{12}^* \Delta Q_{12}, \tag{3.24}$$

$$\frac{\partial \Delta Q_{12}}{\partial t} = 2c_1 M_2^* \Delta M_1 + 2c_1 M_1^* \Delta M_2 - 2Q_{11}^* Q_{12}^* \Delta Q_{11}$$
$$+ \left(\pm 1 - 3Q_{12}^{*2} - Q_{11}^{*2} - \mathbf{k}^2\right) \Delta Q_{12}, \tag{3.25}$$

$$\frac{\partial \Delta M_1}{\partial t} = \left(\pm 1 - 3M_1^{*2} - M_2^{*2} + c_2 Q_{11}^* - \mathbf{k}^2\right) \Delta M_1 + \left(c_2 Q_{12}^* - 2M_1^* M_2^*\right) \Delta M_2$$
$$+ c_2 M_1^* \Delta Q_{11} + c_2 M_2^* \Delta Q_{12}, \tag{3.26}$$

$$\frac{\partial \Delta M_2}{\partial t} = \left(c_2 Q_{12}^* - 2M_1^* M_2^*\right) \Delta M_1 + \left(\pm 1 - 3M_2^{*2} - M_1^{*2} - c_2 Q_{11}^* - \mathbf{k}^2\right) \Delta M_2$$
$$- c_2 M_2^* \Delta Q_{11} + c_2 M_1^* \Delta Q_{12}. \tag{3.27}$$

A solution is considered stable (unstable) if fluctuations decrease (increase) over time. To begin with, the stability of the disordered solution, where $Q_{11}^* = Q_{12}^* = M_1^* = M_2^* = 0$, is examined. The stability characteristics match those of the uncoupled case ($c_1 = c_2 = 0$), since the coupling terms do not affect the leading order. Thus, growth in the **M**-field (for $T < T_c^M$) cannot destabilize the **Q**-field (for $T > T_c^N$) and vice versa in the linearized equations, though it may happen in the fully nonlinear equations.

Next, the stability analysis is presented for a typical non-trivial solution. Consider the evaluation for a prototypical Case 2(ii) where $T_c^M < T < T_c^N$ and $c_1 = 0, c_2 \neq 0$. Then, Eqs. (3.20)–(3.23) take the form:

$$Q_{11}^* - |\mathbf{Q}^*|^2 Q_{11}^* = 0, \tag{3.28}$$
$$Q_{12}^* - |\mathbf{Q}^*|^2 Q_{12}^* = 0, \tag{3.29}$$
$$-M_1^* - |\mathbf{M}^*|^2 M_1^* + c_2 \left(Q_{11}^* M_1^* + Q_{12}^* M_2^*\right) = 0, \tag{3.30}$$
$$-M_2^* - |\mathbf{M}^*|^2 M_2^* + c_2 \left(Q_{12}^* M_1^* - Q_{11}^* M_2^*\right) = 0. \tag{3.31}$$

The non-trivial solution of these equations is given by:

$$Q_{11}^* = \cos 2\theta, \quad Q_{12}^* = \sin 2\theta; \quad M_1^* = r_M \cos \theta, \quad M_2^* = r_M \sin \theta; \tag{3.32}$$

with θ being an arbitrary angle between **M** and x-axis and $r_M = \sqrt{c_2 - 1}$. With the form of the solutions, it is easy to see that the director **n** also makes an angle θ with the x-axis. Thus, in the stationary solution **n** and **M** are co-aligned as is intuitively expected. This characteristic is common in all cases and has significant implications for the nonequilibrium behavior of FNs.

The subsequent task is to evaluate the stability of the stationary solution described in Eq. (3.32). $\theta = 0$ can be chosen without loss of generality given the nature of the solution. The fluctuations in the Fourier components, from Eqs. (3.24)–(3.27) for $c_1 = 0$ and $c_2 \neq 0$, behave as follows:

$$\Delta Q_{11}(\mathbf{k}, t) = \Delta Q_{11}(\mathbf{k}, 0) e^{-(2+k^2)t}, \tag{3.33}$$

$$\Delta Q_{12}(\mathbf{k}, t) = \Delta Q_{12}(\mathbf{k}, 0) e^{-k^2 t}, \tag{3.34}$$

$$\Delta M_1(\mathbf{k}, t) = \left[\Delta M_1(\mathbf{k}, 0) - c_2' \Delta Q_{11}(\mathbf{k}, 0)\right] e^{(2-2c_2-k^2)t}$$
$$+ c_2' \Delta Q_{11}(\mathbf{k}, 0) e^{-(2+k^2)t}, \tag{3.35}$$

$$\Delta M_2(\mathbf{k}, t) = \left[\Delta M_2(\mathbf{k}, 0) - c_2'' \Delta Q_{12}(\mathbf{k}, 0)\right] e^{-(2c_2+k^2)t}$$
$$+ c_2'' \Delta Q_{12}(\mathbf{k}, 0) e^{-k^2 t}, \tag{3.36}$$

where $c_2' = (c_2\sqrt{c_2-1})/(2c_2-4)$ and $c_2'' = (\sqrt{c_2-1})/2$. It is clear that for any arbitrary $\mathbf{k}$, the fluctuations in $\mathbf{Q}$ consistently decrease, indicating that the solution is stable. However, ΔM_1 is stable only if $c_2 > 1$. Linear stability analysis can often establish constraints on c_1 and c_2, which guide their selection in simulations. The stationary solutions are tabulated for all cases in Tables A.1–A.3 of Appendix A.

3.2.4 Morphology Characterization

A commonly used probe to understand evolving morphologies is the spatial correlation function defined in terms of the order parameter field $\boldsymbol{\psi}(\mathbf{r}, t)$ as [1]:

$$C(\mathbf{r}, t) = \frac{1}{V} \int d\mathbf{R} \left[\langle \boldsymbol{\psi}(\mathbf{R}, t) \cdot \boldsymbol{\psi}(\mathbf{R} + \mathbf{r}, t) \rangle - \langle \boldsymbol{\psi}(\mathbf{R}, t) \rangle \cdot \langle \boldsymbol{\psi}(\mathbf{R} + \mathbf{r}, t) \rangle\right], \tag{3.37}$$

where V is the volume of the system and $\langle \cdots \rangle$ indicates an averaging over independent runs. Small-angle scattering experiments measure it's Fourier transform which is the equal-time structure factor:

$$S(\mathbf{k}, t) = \int d\mathbf{r} \, e^{i\mathbf{k} \cdot \mathbf{r}} C(\mathbf{r}, t), \tag{3.38}$$

where $\mathbf{k}$ is the wave vector of the the scattered beam. Isotropic systems are characterized by a single length scale $L(t)$.

The correlation function in coarsening systems exhibits a dynamical scaling form [1]:

$$C(\mathbf{r}, t) = f\left(\frac{r}{L}\right), \tag{3.39}$$

where $f(x)$ is the scaling function. The characteristic length scale $L(t)$ is defined as the distance over which the correlation function decays to (say) half of its maximum value. Typically, it represents the average distance between two defects. The corresponding dynamical scaling form for the structure factor is given by

$$S(\mathbf{k}, t) = L^d g(kL), \tag{3.40}$$

3.2 Theoretical Framework

where $g(p)$ is the Fourier transform of $f(x)$. The characteristic length can also be calculated from the structure factor as the inverse of its first moment [1]:

$$L(t) = \left[\frac{\int \mathbf{k}\, S(\mathbf{k}, t)\, d\mathbf{k}}{\int S(\mathbf{k}, t)\, d\mathbf{k}}\right]^{-1}, \qquad (3.41)$$

An approximate analytical form of the correlation function for a general n-component system with non-conserved dynamics has been obtained by Bray and Puri [30] and Toyoki [31] by studying the defect dynamics. The Bray-Puri-Toyoki (BPT) function is valid for the systems with topological defects when $n \leq d$ and has the following analytical form:

$$C(r, t) = \frac{n\gamma}{2\pi} \left[B\left(\frac{n+1}{2}, \frac{1}{2}\right)\right]^2 F\left(\frac{1}{2}, \frac{1}{2}; \frac{n+2}{2}; \gamma^2\right). \qquad (3.42)$$

In the above expression the beta function $B(x, y) = \Gamma(x)\Gamma(y)/\Gamma(x+y)$, $F(a, b; c; z)$ is a hyper-geometric function [32], and $\gamma = \exp\left[-r^2/(2L^2)\right]$. BP also demonstrated that the corresponding scaling function

$$g(p) \sim p^{-(d+n)} \quad \text{for} \quad p \to \infty. \qquad (3.43)$$

This outcome is termed the *generalized Porod law*. The structure factor tail indicates the prevalence of several types of topological defects inside the system [30]. For $n = 1$, the defects are interfaces, and the associated scattering function demonstrates the *Porod law* [33, 34]. For $n > 1$, the distinct topological defects include vortices ($n = 2, d = 2$), strings ($n = 2, d = 3$), and monopoles or hedgehogs ($n = 3, d = 3$). In $d = 3$, the behavior of $g(p)$ behaving as p^{-5} or p^{-6} is contingent upon whether strings or monopoles predominate.

Analyzing the growth law of domains [$L(t)$ versus t] is a crucial element in coarsening experiments, as it provides insight into the free energy landscape and the relaxation time scales within the system. For instance, in pure isotropic systems exhibiting non-conserved dynamics, the growth follows the Lifshitz–Allen–Cahn (LAC) law where $L(t) \sim t^{1/2}$ [35]. In contrast, pure isotropic systems with conserved dynamics and diffusive transport adhere to the Lifshitz–Slyozov (LS) law characterized by $L(t) \sim t^{1/3}$ [36]. These growth laws are indicative of systems without energy barriers to coarsening, which exhibit a single relaxation time scale. The evolution of the order parameter ψ in a non-conserved system, characterized by a scalar order parameter, is described by the TDGL equation: $\partial \psi/\partial t = \psi - \psi^3 + \nabla^2 \psi$, written in dimensionless units [1]. Given that ψ exhibits a *kink* profile at the domain interface, the equation of motion can be derived by reformulating the TDGL equation in terms of interfacial coordinates [35]. This leads to the Allen–Cahn equation $dL/dt = -(d-1)/L$, which, upon integration, produces the LAC law $L(t) \sim t^{1/2}$.

In the interest of completeness, we also explore the growth law in systems with conserved order parameters, such as the kinetics of phase separation in a binary (AB) mixture. In this scenario, an appropriate order parameter is the density difference between the two species: $\psi = n_A - n_B$. For systems characterized solely by diffusive transport (solid mixtures), the evolution of the order parameter is governed by the Cahn-Hilliard (CH) equation:

$$\frac{\partial \psi}{\partial t} = \nabla^2 \mu = \nabla^2 [-\psi + \psi^3 - \nabla^2 \psi], \qquad (3.44)$$

where the units are also dimensionless [1]. Similar to the TDGL equation, it is assumed that the interface between domains is in local equilibrium. This assumption enables the derivation of the conserved counterpart of the Allen–Cahn equation. However, this process is significantly more complex because the conservation constraint means that the velocity at any point on the interface must satisfy an integral equation that involves all other interfaces in the system [2]. Despite this complexity, the resulting equation can still be analyzed dimensionally. The chemical potential at the surface of a domain with size L is given by $\mu \sim \sigma/L$, where σ represents the surface tension. The corresponding current is $j = |\nabla \mu| \sim \sigma/L^2$. The domain size grows as $dL/dt \sim \sigma/L^2$, which on integrating yields the LS law $L(t) \sim (\sigma t)^{1/3}$.

3.3 Detailed Numerical Results

All simulations in $d = 2$ FN have been performed on a system of size $N^2 = (2048)^2$, while the results in $d = 3$ have been obtained for $N^3 = (256)^3$. Periodic boundary conditions are applied to mitigate edge effects. The initial values of the components of **Q** and **M** consist of small fluctuations around zero, simulating a disordered system prior to the quench. The evolution is studied by numerically solving Eqs. (3.6)–(3.9) and (3.11)–(3.18) using the Euler discretization method [37]. The spatial and temporal scales for the system are defined by the discretization mesh sizes Δx and Δt. The impact of these mesh size choices is discussed in Sect. 2.3. Before proceeding, it is important to address the influence of system size on the simulations. Previous studies on domain growth [38–40] indicate that finite-size effects become significant when the characteristic domain scale exceeds approximately 25% of the lateral system size. These effects manifest as a slowdown in the domain growth law, leading to an underestimation of the growth exponent on a log-log scale. To avoid artifacts resulting from system size, results are presented for time windows where $L(t) < (N\Delta x)/4$.

3.3.1 Domain Growth in $d = 2$ Ferronematics

The $d = 2$ simulations mimic square-well experiments in which the height of the sample is much smaller than the emergent length scale of the emergent morphologies. This geometry has generally been realized in experiments by surface treatment of

3.3 Detailed Numerical Results

the top and bottom layers of the sample [41], and more recently through confinement between two substrates [42]. Equations (3.6)–(3.9) are solved on a square lattice of size $(N\Delta x)^2$. The mesh sizes are $\Delta x = 1$ and $\Delta t = 0.01$. The orientation of the director can be obtained from the Q_{ij}'s using Eq. (2.1). The nematic and magnetic morphologies represent the orientation of **n** and **M** at each point on the square grid at the given time. The magnitudes agree with those of the stationary solutions, except at the defects where the respective order parameters are $\simeq 0$. The numerical results are obtained for all the cases in Table 3.1, and some representative results are presented below. The results for other cases are summarized in Tables A.1–A.3 and 3.2.

The parameters c_1 and c_2 are assigned representative values for the numerical results presented here. As mentioned above, c_1 denotes the coupling effect on the nematic component resulting from the magnetic component, while c_2 measures the coupling effect on the magnetic component induced by the nematic component. These represent various limiting cases that are relevant to experimental contexts. Universal results, such as growth law exponents and scaling functions, remain invariant with respect to the selection of parameters. The parameter values influence only non-universal quantities, such as prefactors or time scales of growth laws, which are of comparatively lesser significance. This aligns with our understanding of static critical phenomena.

In Fig. 3.2, the left side illustrates the nematic morphology, while the right side shows the magnetic structure at $t = 10^3$. Figure 3.2a, b represent the system for a quench temperature where $T < \min\{T_c^N, T_c^M\}$ with $c_1 = c_2 = 0$, which corresponds to the uncoupled case. It is important to remember that the nematic director **n** has an inversion symmetry. As a result, in the nematic images, the blue regions signify **n** pointing in the first (or third) quadrant, while the green regions correspond to **n** pointing in the second (or fourth) quadrant. In the magnetization snapshots, four distinct colors indicate **M** lying in the first, second, third, and fourth quadrants, respectively. Since there is no magneto-nematic coupling, the two structures evolve independently.

Figure 3.2c, d correspond to Case 1(i), where $c_1 = 4$ and $c_2 = 0$, using the same initial configuration as in Fig. 3.2a, b. Given that $T_c^N < T < T_c^M$, linear stability analysis predicts an isotropic phase for the nematic component and a ferromagnetic phase for the magnetic component. A notable phenomenon here is the *slaved coarsening* of the nematic phase, which occurs due to its interaction with the ordered magnetic component. The **M**-field, being unstable, grows over time, and once it reaches a significant magnitude, it pulls the **Q**-field into the growth process. This behavior is strictly a non-linear effect, as the linearized equations are identical to those in the uncoupled case where $c_1 = c_2 = 0$. As seen in Fig. 3.2c, d, the magnetic vector **M** is parallel or anti-parallel to **n**, a result of the coupling term in the free energy described by Eq. (3.1). Furthermore, the magnitudes of both **M** and **Q** are consistent with the stationary solutions listed in Table A.1. The spatial variation in **n** and **M**, as shown in Fig. 3.2 and in later snapshots, is determined by the type of defects present in the system. For instance, if the defects are interfaces (as will be demonstrated for sub-

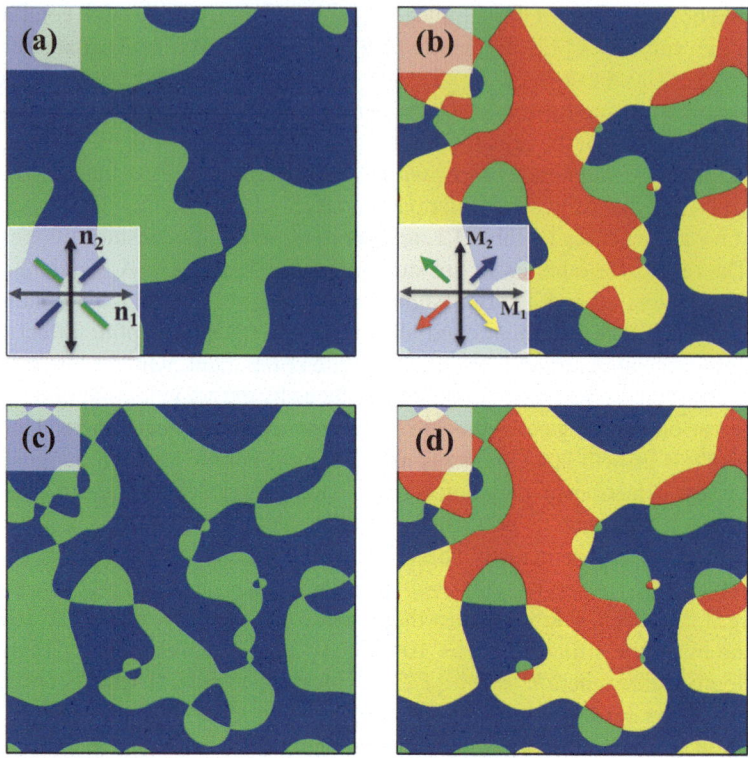

Fig. 3.2 (color online) Domain growth in $d = 2$ ferronematics: nematic morphologies (snapshots on the left) and magnetic morphologies (snapshots on the right) at $t = 10^3$. The frames **a**, **b** correspond to the uncoupled case, i.e., $c_1 = c_2 = 0$ with quench temperature $T < \min\{T_c^N, T_c^M\}$. The frames **c**, **d** correspond to Case 1(i), i.e., $c_1 = 4$, $c_2 = 0$ of Table 3.1. The insets in **a**, **b** show the color scheme used to depict the order parameters. Reprinted with permission from [17]. © 2021, Royal Society of Chemistry. All rights reserved

domain morphologies or SDM), the domain walls are sharp and narrow. However, if the defects are vortices or strings, the order parameters exhibit smooth variations across space.

The characteristic features of the domains and defects are evaluated by computing $C(r, t)$ and $S(k, t)$. The magnetic correlation function $C_M(r, t)$ is calculated directly from the definition in Eq. (3.37), with ψ replaced by **M**. For the nematic component, $C_Q(r, t)$ is obtained from the tensor order parameter **Q**. The correlated regions in the nematic and magnetic components are characterized by the two length scales L_Q and L_M respectively. They are defined as the distance at which the correlation function decays to half its maximum value. Figure 3.3a, b present the data for $C_Q(r, t)$ versus $r/L_Q(t)$ and $C_M(r, t)$ versus $r/L_M(t)$ for Case 1(i) with $c_1 = 4$ and $c_2 = 0$ at times $t = 10^2, 10^3, 10^4$. Both sets of data show excellent collapse, indicating dynamical scaling. This suggests that the morphologies of the nematic

3.3 Detailed Numerical Results

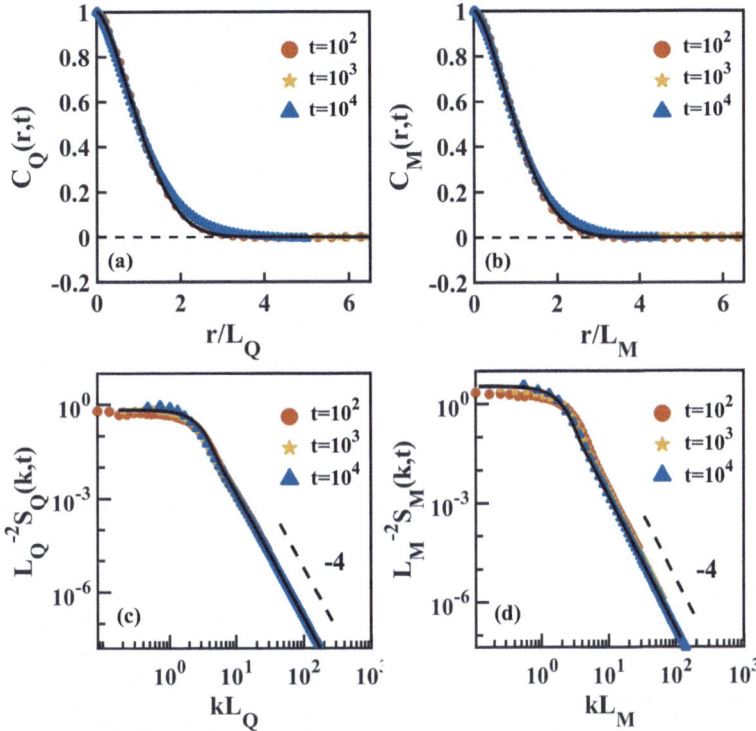

Fig. 3.3 Dynamical scaling of correlation functions and structure factors in $d = 2$. The data for the nematic (left) and magnetic (right) components are shown for Case 1(i), i.e., $c_1 = 4$, $c_2 = 0$. The solid lines in the frames denote the BPT function [see Eq. (3.42)] evaluated for $n = 2, d = 2$. Reprinted with permission from [17]. © 2021, Royal Society of Chemistry. All rights reserved

and magnetic domains in the coupled system remain unchanged over time except for a scale factor. The solid line in Fig. 3.3a, b represents the BPT function calculated using $n = 2$, $d = 2$, and matches well with the data. Figure 3.3c, d show the structure factors $L_Q^{-2} S_Q(k, t)$ versus kL_Q and $L_M^{-2} S_M(k, t)$ versus kL_M. Both components exhibit asymptotic tails with behavior k^{-4}, indicating the presence of vortex defects during the ordering process. In fact, $S(k) \sim k^{-4}$ is observed whenever (i) $c_1 \neq 0, c_2 = 0$ and (iii) $c_1 = c_2 = c$, as noted in Table 3.2. Since it is unlikely that c_1 or c_2 would be exactly 0 in the experimental systems, the robustness of these results with respect to deviations from the limiting cases (i)–(iii) in Table 3.1 is examined. Figure 3.4 displays the growth law $L(t)$ versus t on a log-log scale for nematic and magnetic components for Case 1 with $c_1 = 3$ and $c_2 = 0, 1, 2$. On this scale, the data for both components show no significant deviation from the $c_2 = 0$ case even for values of c_2 approaching 1. This confirms that the results for the limiting case (i) remain valid when $c_1 \gg c_2$. Similar conclusions apply when $c_2 \gg c_1$.

Table 3.2 Quench temperature, coupling limits, growth laws and structure factor tails for nematic and magnetic components in $d = 2$

Quench temperature	Coupling limits	Growth laws (L_Q, L_M)	Tails of $S(k,t)$ (S_Q, S_M)
Case 1 $(T_c^N < T < T_c^M)$	(i) $c_1 \neq 0$, $c_2 = 0$ (ii) $c_1 = 0$, $c_2 \neq 0$ (iii) $c_1 = c_2 = c$	$[(t/\ln t)^{1/2}, (t/\ln t)^{1/2}]$ [No growth, $(t/\ln t)^{1/2}$] $[(t/\ln t)^{1/2}, (t/\ln t)^{1/2}]$	(k^{-4}, k^{-4}) (Uniform, k^{-4}) (k^{-4}, k^{-4})
Case 2 $(T_c^M < T < T_c^N)$	(i) $c_1 \neq 0$, $c_2 = 0$ (ii) $c_1 = 0$, $c_2 \neq 0$ (iii) $c_1 = c_2 = c$	$[(t/\ln t)^{1/2}, \text{No growth}]$ $[(t/\ln t)^{1/2}, \text{Saturation}]$ $[(t/\ln t)^{1/2}, (t/\ln t)^{1/2}]$	$(k^{-4}, \text{Uniform})$ (k^{-4}, k^{-3}) (k^{-4}, k^{-4})
Case 3 $(T < \min\{T_c^N, T_c^M\})$	(i) $c_1 \neq 0$, $c_2 = 0$ (ii) $c_1 = 0$, $c_2 \neq 0$ (iii) $c_1 = c_2 = c$	$[(t/\ln t)^{1/2}, (t/\ln t)^{1/2}]$ $[(t/\ln t)^{1/2}, \text{Saturation}]$ $[(t/\ln t)^{1/2}, (t/\ln t)^{1/2}]$	(k^{-4}, k^{-4}) (k^{-4}, k^{-3}) (k^{-4}, k^{-4})

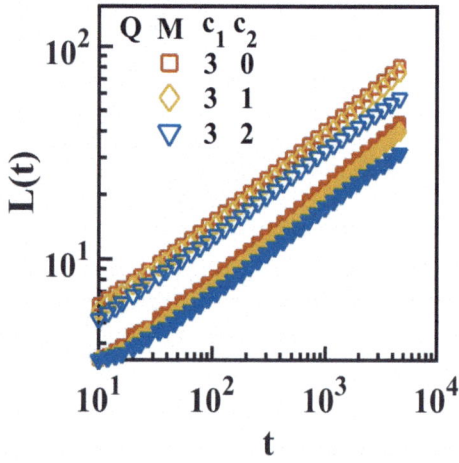

Fig. 3.4 Growth laws in $d = 2$ (on a log-log scale) for the nematic (solid symbols) and magnetic (open symbols) components for Case 1 in Table 3.1 with $c_1 = 3$ and $c_2 = 0, 1, 2$. Reprinted with permission from [17]. © 2021, Royal Society of Chemistry. All rights reserved

The analysis of growth laws in this coupled system is considered next. Figure 3.5 presents the length scale data for two cases: (a) Case 1(i), where $c_1 = 3, 4, 5$ and $c_2 = 0$; and (b) Case 1(iii), where $c_1 = c_2 = c$ for $c = 3, 4, 5$. Solid symbols represent nematic data, while open symbols correspond to the magnetic component. For $T_c^N < T < T_c^M$, the nematic component is naturally disordered but is forced to order due to the magneto-nematic coupling. In the uncoupled limit, the growth law for both fields shows slower than $t^{1/2}$ growth due to logarithmic corrections that lead to $L(t) \sim (t/\ln t)^{1/2}$ [1, 2]. It is evident that the nematic length scale L_Q remains smaller than the magnetic length scale L_M at all times, indicating that the nematic field is influenced by the magnetic field. Interestingly, the growth laws do not show any significant variation with respect to the strength of the coupling as it does not change the nature of the coarsening process. To quantify these growth laws, the

3.3 Detailed Numerical Results

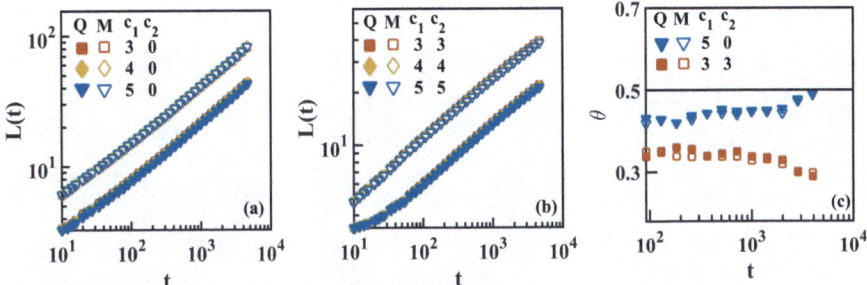

Fig. 3.5 Log-log plot of growth laws in $d = 2$ for the nematic (solid symbols) and magnetic (open symbols) components for **a** Case 1(i), and **b** Case 1(iii). **c** Effective growth exponent $\theta = d(\ln L)/d(\ln t)$ versus t on a log-linear scale. The solid line corresponds to $\theta = 1/2$. The symbol sizes are larger than the error bars. Reprinted with permission from [17]. © 2021, Royal Society of Chemistry. All rights reserved

effective exponent $\theta = d \ln L(t)/d \ln t$ is calculated, as shown in Fig. 3.5c on a semi-log plot. Since θ is derived from a discrete derivative, some noise is expected in the data. Note that θ does not remain constant over time, but rather follows a power-law approximation of the growth law $L(t) \sim (t/\ln t)^{1/2}$. For the case $c_1 = c_2 = c$, the value of θ is approximately 0.35.

The next scenario explores the impact of the freely evolving nematic component on the magnetic field, focusing on the case where $c_1 = 0$ and $c_2 \neq 0$. Consider Case 2(ii) corresponding to the quench temperature $T_c^M < T < T_c^N$. Here, the linear stability analysis indicates that the **Q**-field becomes unstable, while the disordered **M**-field remains stable. The stationary solutions for this case, discussed in Sect. 3.2.2 (see Eq. (3.32)), demonstrate that non-trivial solutions for **M** arise only when $c_2 > 1$. Figure 3.6 illustrates the time evolution of the nematic (upper row) and magnetic (lower row) structures for $c_1 = 0$ and $c_2 = 4$ at $t = 10^2$, $t = 10^3$ and $t = 10^4$. Since the magnetic component is linearly stable, the nematic field must first grow to a sufficient magnitude to influence the magnetic field. Interestingly, the magnetic component reveals a sub-domain morphology (SDM) characterized by two distinct alignments: $\mathbf{n} \parallel \mathbf{M}$ and $\mathbf{n} \parallel -\mathbf{M}$. Although these structures incur a surface tension cost, the entropic gain is large enough to stabilize the morphologies. Furthermore, the magnitudes of **Q** and **M** are consistent with the stationary solutions from Eq. (3.32). Interestingly, similar SDMs have been observed experimentally in ferromagnetic nematics when the system is quenched in the absence of the external field [22].

Figure 3.7a shows the growth behavior for $c_2 = 3$ and 4. The nematic field length scale follows a growth law $L_Q(t) \sim t^\theta$, with $\theta < 0.5$ due to the presence of logarithmic corrections. Meanwhile, the magnetic field length scale, $L_M(t)$, reaches a steady state because of the formation of SDMs. In Fig. 3.7b, the relationship between the saturated length scale L_M^S and the coupling constant c_2 is displayed for a $(2048)^2$

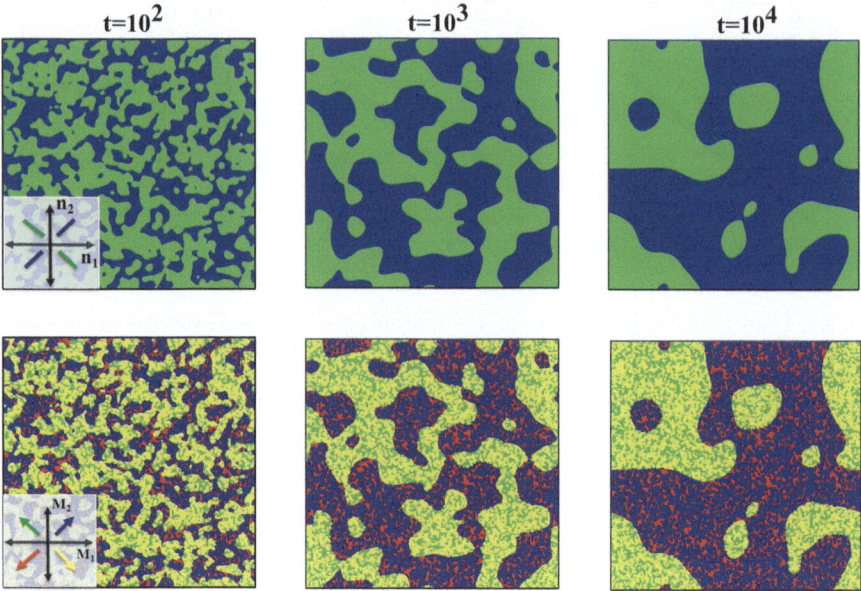

Fig. 3.6 Evolution morphologies in $d = 2$ for Case 2 (ii), i.e., $c_1 = 0$, $c_2 = 4$. The snapshots at $t = 10^2, 10^3, 10^4$ for the nematic (top) and magnetic (bottom) components are shown. Reprinted with permission from [17]. © 2021, Royal Society of Chemistry. All rights reserved

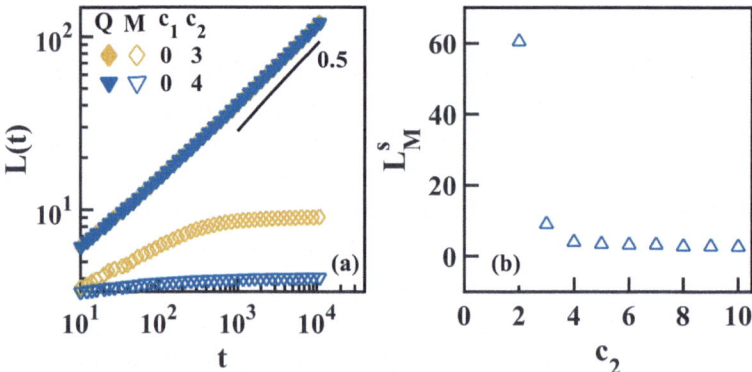

Fig. 3.7 Domain growth in $d = 2$ for Case 2(ii). **a** Growth laws for the nematic (solid symbols) and magnetic (open symbols) components. **b** Saturation length L_M^s for different values of c_2. Reprinted with permission from [17]. © 2021, Royal Society of Chemistry. All rights reserved

3.3 Detailed Numerical Results

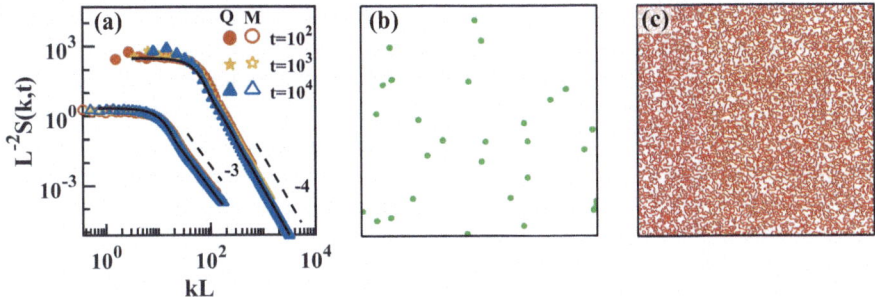

Fig. 3.8 (a) Scaled structure factors for nematic (solid symbols) and magnetic (open symbols) components for Case 2(ii) with $c_1 = 0$, $c_2 = 4$. The data sets have been shifted for clarity. The solid lines in (a) denote the BPT function calculated for $n = 1$, $d = 2$ (M) and $n = 2$, $d = 2$ (Q). Defect locations of the (b) **Q**-field and (c) **M**-field for morphologies at $t = 10^3$ in Fig. 3.6. The defects are defined as regions where $S < 0.6$ and $|\mathbf{M}| < 0.6$, respectively. Reprinted with permission from [17]. © 2021, Royal Society of Chemistry. All rights reserved

system. Although the exact dependence of L_M^S on c_2 remains unclear, it is expected that L_M^S decreases to zero as $c_2 \to \infty$ and increases indefinitely as c_2 approaches 1.

Further insights into the SDM are gained by examining the scaled structure factor $L^{-2}S(k, t)$ versus kL, as shown in Fig. 3.8a for $c_1 = 0$, $c_2 = 4$ and $t = 10^2, 10^3, 10^4$. The data for both the nematic and magnetic fields collapse well, indicating dynamical scaling. The solid lines depicting the Fourier transform of the BPT function with parameters $n = 2, d = 2$ for the nematic component and $n = 1, d = 2$ for the magnetic component align neatly with the numerical data. Moreover, the tail of the structure factor follows $S_Q(k, t) \sim k^{-4}$, which indicates a generalized Porod tail driven by vortex defects in the nematic structure. In contrast, the magnetic structure factor exhibits $S_M(k, t) \sim k^{-3}$, indicative of interfacial scattering despite the continuous nature of the magnetic order parameter. To confirm these findings, Fig. 3.8b and c highlight the defect locations for the nematic and magnetic fields, corresponding to the morphology at $t = 10^3$ from Fig. 3.6. Defects are identified as regions where $S < 0.6$ and $|\mathbf{M}| < 0.6$. The figures depict the isolated vortex defects in the nematic field and prominent interfacial defects in the magnetic field. These patterns are typical in the limit $c_1 = 0$, $c_2 \neq 0$, as seen in Table 3.2 regardless of the quench temperature. A key application of this observation is that the coupling strength can control the characteristic length scale in SDMs, thereby offering the potential to design patterns in FNs.

3.3.2 Domain Growth in $d = 3$ Ferronematics

In $d = 3$, the **Q**-tensor is a symmetric and traceless 3×3 matrix, and $\text{Tr}(\mathbf{Q})^3 \neq 0$. This introduces a cubic term in the GL free energy [see Eq. (3.1)], resulting in a first-order nematic-isotropic transition, in contrast to the continuous transition observed in $d = 2$. Additionally, the $d = 3$ setting permits biaxiality, leading to more complex defect structures such as strings and hedgehogs. The **Q**-tensor has five degrees of freedom ($\{q_i\}$, $i = 1, 2, \ldots, 5$), while the magnetization **M** has three components ($(\{M_j\}$, $j = 1, 2, 3$). The set of coupled equations governs the temporal evolution of these eight variables [Eqs. (3.11)–(3.18)], in their dimensionless form. For numerical stability, the mesh sizes are $\Delta x = 1$ and $\Delta t = 0.02$. These equations determine the values of the q_i's and M_j's at every lattice point. The **Q**-tensor remains symmetric and traceless, although not necessarily diagonal. Thus, a reference frame aligned with its principal axis is selected to diagonalize the **Q**-tensor. The largest eigenvalue provides $\mathcal{S}$ [cf. Eq. (2.1)], and the corresponding eigenvector represents **n**. The parameters c_1 and c_2 are similar to $d = 2$, with c_1 representing the influence of **M** on **Q**, and c_2 characterizing the effect of **Q** on **M**.

Figure 3.9 shows the evolving patterns for both the nematic (top row) and magnetic (second row) components, corresponding to Case 1(i) in Table 3.1, where $c_1 = 3$ and $c_2 = 0$. In this scenario, the nematic field (**Q**) exhibits coarsening behavior

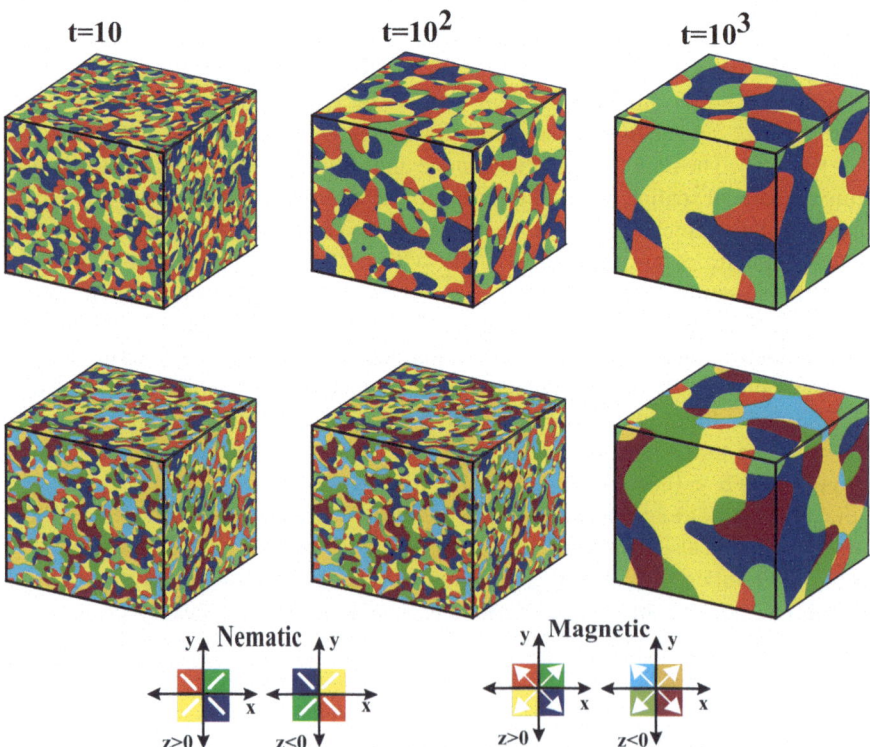

Fig. 3.9 Domain growth for $d = 3$ FNs. Nematic (top row) and magnetic (bottom row) morphologies for Case 1(i) with $c_1 = 3$, $c_2 = 0$. Snapshots at $t = 10$, 10^2, 10^3 are shown. The color code used for the snapshots is shown at the bottom. Reprinted with permission from [17]. © 2021, Royal Society of Chemistry. All rights reserved

3.3 Detailed Numerical Results

influenced by the magnetic field (**M**). The color coding is illustrated in the bottom row of the figure. Due to the symmetry of the nematic field, four distinct colors are used for the director field **n**, while the magnetic field, lacking this symmetry, is represented using eight colors. Despite the temperature relationship $T_c^N < T < T_c^M$, **n** aligns with **M** or $-\mathbf{M}$ due to the magneto-nematic coupling. To analyze the morphologies shown in Fig. 3.9, the correlation function and the structure factor are calculated. The nematic correlation function, $C_Q(r, t)$, is obtained from Eq. (3.37) by substituting $\psi = P_2(\cos\theta)$. Similarly, the magnetic correlation function, $C_M(r, t)$, is evaluated following the same approach as previously. In Fig. 3.10a, the plot of $C_Q(r, t)$ versus scaled distance $r/L(t)$ is shown for the case with coupling strength $c_1 = 3$, $c_2 = 0$, at times $t = 100, 300, 500$ for the nematic field (represented by solid symbols). Meanwhile, Fig. 3.10b illustrates the corresponding correlation data

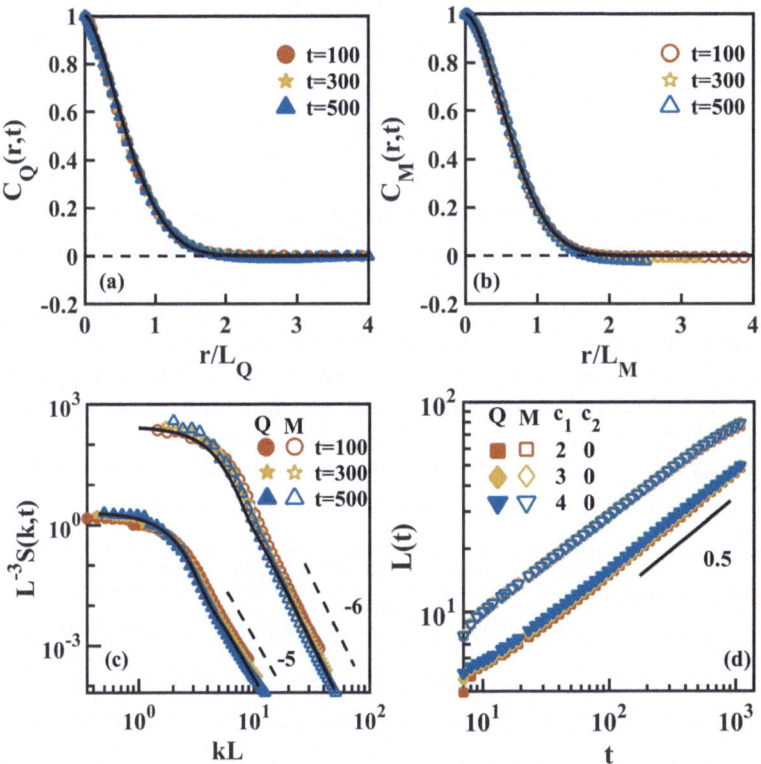

Fig. 3.10 Dynamical scaling of the correlation function for the evolution in Fig. 3.9: **a** nematic component, and **b** magnetic component. The BPT function (solid line) in (**a**) is evaluated for $n = 2$, $d = 3$, and in (**b**) for $n = 3$, $d = 3$. **c** Scaled structure factors for the nematic (solid symbols) and magnetic (open symbols) components. The corresponding BPT functions are shown as solid lines. The data sets have been shifted for clarity. **d** Growth laws for $c_2 = 0$ and $c_1 = 2, 3, 4$. Reprinted with permission from [17]. © 2021, Royal Society of Chemistry. All rights reserved

for the magnetic field (indicated by open symbols). The excellent data collapse in both figures demonstrates dynamical scaling for the evolution. $C_Q(r, t)$ data fit well to the BPT function calculated for $n = 2$ in contrast to the data for $C_M(r, t)$ which is described well by the BPT function with $n = 3$. This highlights a clear difference in behavior at small r/L, leading to distinct defect structures in the two components, which will be discussed further. The related structure factor plotted as $L^{-3}S(k, t)$ versus kL in Fig. 3.10c shows data that have been offset for clarity. Consistent with expectations, the generalized Porod law $S_M(k, t) \sim k^{-6}$ ($d = 3, n = 3$) describes the scattering from monopole defects in the magnetic field. On the other hand, $S_Q(k, t) \sim k^{-5}$, indicating that $+1/2$ string defects dominate the nematic configurations. The variation in the large k behavior between S_Q and S_M is due to the different small r behavior noted in Fig. 3.10a, b. In Fig. 3.10d, the length scale $L(t)$ versus t is plotted for both fields on a log-log scale for $c_1 = 2, 3, 4$. The nematic field exhibits a length scale smaller than that of the magnetization field because of the delayed response caused by the coupling. The data is well-represented by the LAC law $L(t) \sim t^{1/2}$, and no logarithmic corrections are present in $d = 3$. Table 3.3 summarizes the observations corresponding to other cases.

Lastly, a set of exotic morphologies arising from Case 2 (ii) are shown in Fig. 3.11a, b for $c_1 = 0, c_2 = 3$ at $t = 500$. The coupling between **n** and **M** leads to magnetic ordering, but the resulting patterns show SDM akin to those seen in $d = 2$ FNs. Figure 3.12a shows the growth behavior of the nematic **Q**-field and magnetic **M**-field (solid and open symbols, respectively) for Case 2 (ii). As anticipated, $L_Q(t) \sim t^{1/2}$, but $L_M(t)$ stabilizes to L_M^s over time due to SDM formation. Figure 3.12b presents scaled structure factor plots for $c_1 = 0, c_2 = 3$, both exhibiting dynamical scaling, indicating the presence of a dominant length scale. The nematic component reveals a generalized Porod tail, $S_Q(k, t) \sim k^{-5}$, suggesting string defects characteristic of $d = 3, n = 2$. However, the magnetic component follows the standard Porod law, $S_M(k, t) \sim k^{-4}$, indicative of sharp interface scattering between regions with **M** and $-\mathbf{M}$, typical of $d = 3, n = 1$. The fitting to BPT functions for $n = 2$ and $n = 1$

Table 3.3 Quench temperature, coupling limits, growth laws and structure factor tails for nematic and magnetic components in $d = 3$

Quench temperature	Coupling limits	Growth laws (L_Q, L_M)	Tails of $S(k, t)$ (S_Q, S_M)
Case 1 $(T_c^N < T < T_c^M)$	(i) $c_1 \neq 0$, $c_2 = 0$ (ii) $c_1 = 0$, $c_2 \neq 0$ (iii) $c_1 = c_2 = c$	$(t^{1/2}, t^{1/2})$ (No growth, $t^{1/2}$) $(t^{1/2}, t^{1/2})$	(k^{-5}, k^{-6}) (Uniform, k^{-6}) (k^{-5}, k^{-6})
Case 2 $(T_c^M < T < T_c^N)$	(i) $c_1 \neq 0$, $c_2 = 0$ (ii) $c_1 = 0$, $c_2 \neq 0$ (iii) $c_1 = c_2 = c$	$(t^{1/2}$, No growth)) $(t^{1/2}$, Saturation) $(t^{1/2}, t^{1/2})$	$(k^{-5}$, Uniform) (k^{-5}, k^{-4}) (k^{-5}, k^{-6})
Case 3 $(T < \min\{T_c^N, T_c^M\})$	(i) $c_1 \neq 0$, $c_2 = 0$ (ii) $c_1 = 0$, $c_2 \neq 0$ (iii) $c_1 = c_2 = c$	$(t^{1/2}, t^{1/2})$ $(t^{1/2}$, Saturation) $(t^{1/2}, t^{1/2})$	(k^{-5}, k^{-6}) (k^{-5}, k^{-4}) (k^{-5}, k^{-6})

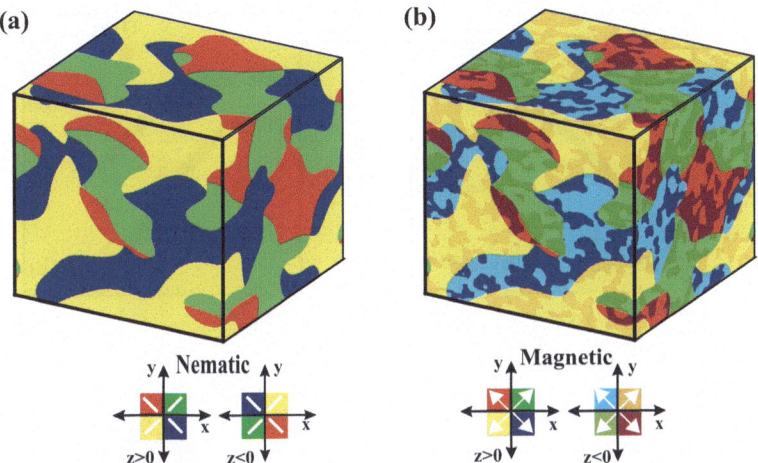

Fig. 3.11 **a** Nematic and **b** magnetic snapshots in $d = 3$ for Case 2(ii), i.e., $c_1 = 0$, $c_2 = 3$. The snapshots correspond to $t = 500$. Reprinted with permission from [17]. © 2021, Royal Society of Chemistry. All rights reserved

confirms distinct defect structures for nematic and magnetic fields, respectively. Figure 3.12c, d display defect locations in the nematic and magnetic fields at $t = 500$ for $c_1 = 0, c_2 = 3$, highlighting string defects in the nematic field and interfacial defects in the magnetic field, consistent with the behavior seen in the structure factor tails for each case.

3.4 Summary and Discussion

In summary, this chapter provided novel and significant insights into the coarsening behavior of FNs in $d = 2, 3$. This system was initially proposed in the 1970s by Brochard and de Gennes [43]. These authors theoretically suggested the formation of FNs through the inclusion of magnetic particles in NLCs. Their work hypothesized that these systems could exhibit spontaneous magnetization without requiring external magnetic fields. Almost four decades later, this theoretical framework was experimentally validated by Mertelj et al. [22], who succeeded in creating the first stable suspension of FNs. Such suspensions offer a unique platform for investigating magneto-mechanical and magneto-optic effects in NLCs, making them relevant for various applications, including photonics, optical switches, microfluidics, and even cosmological studies. Despite these promising applications, further theoretical work on FNs has remained limited. Our research provides a novel framework for understanding domain growth and coarsening in this two-component system. This work is one of the first discussions of nonequilibrium aspects of phase transitions in FNs.

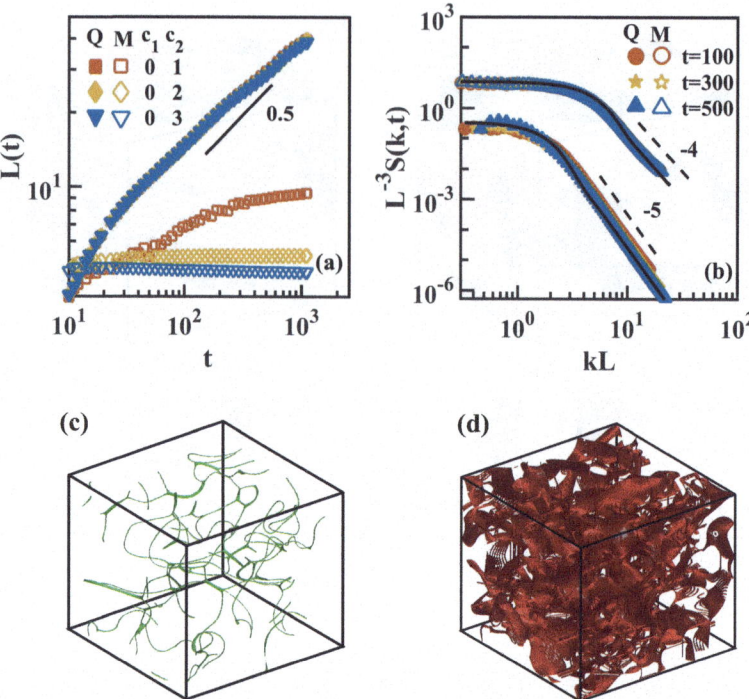

Fig. 3.12 a Growth laws for nematic (solid symbols) and magnetic (open symbols) components for Case 2(ii). **b** Dynamical scaling of the structure factor for $c_1 = 0$, $c_2 = 3$. The data has been shifted for clarity. The solid lines denote the BPT function calculated for $n = 2$, $d = 3$ (Q) and $n = 1$, $d = 3$ (M). Defect regions in **c** nematic component and **d** magnetic component. These are identified as regions where $S < 0.6$ and $|\mathbf{M}| < 0.6$, and are plotted at $t = 500$. Reprinted with permission from [17]. © 2021, Royal Society of Chemistry. All rights reserved

The behavior of FNs is governed by two order parameters: the nematic field is described by the **Q**-tensor, and the magnetization field is described by the vector **M**. The Ginzburg–Landau–de Gennes formalism, incorporating a dyadic magneto-nematic coupling term, provides a framework for modeling the surface interactions between the magnetic particles and NLC molecules. Various temperature quenches were performed in the study to explore the kinetics of phase transitions, utilizing coupled TDGL equations for the nematic and magnetic fields. In this model, two phenomenological parameters play crucial roles: c_1, which quantifies the influence of the nematic order on the magnetic order, and c_2, which characterizes the reverse effect. These coupling constants emerge from the magneto-nematic interaction term in the free energy, and in a dimensionless formulation, they represent the natural scaling of the order parameters. Both parameters can be determined from experimental measurements, and their interplay leads to the formation of stable and exotic morphologies that are absent in systems without coupling. Additionally, the study demonstrated that this coupling mechanism can result in the spontaneous emergence of domain walls and topological defects, even in the absence of external magnetic fields.

References

1. S. Puri, V. Wadhawan, *Kinetics of Phase Transitions* (CRC Press, 2009)
2. A.J. Bray, Adv. Phys. **51**, 481 (2002)
3. R.E. Blundell, A.J. Bray, Phys. Rev. A **46**, R6154 (1992)
4. A.J. Bray, S. Puri, R.E. Blundell, A.M. Somoza, Phys. Rev. E **47**, 2261 (1993)
5. M. Zapotocky, P.M. Goldbart, N. Goldenfeld, Phys. Rev. E **51**, 1216 (1995)
6. C. Denniston, E. Orlandini, J.M. Yeomans, Phys. Rev. E **64**, 021701 (2001)
7. S.M. Kamil, A.K. Bhattacharjee, R. Adhikari, G.I. Menon, Phys. Rev. E **80**, 041705 (2009)
8. A.K. Bhattacharjee, G.I. Menon, R. Adhikari, J. Chem. Phys. **133**, 044112 (2010)
9. A. Singh, S. Ahmad, S. Puri, S. Singh, Europhys. Lett. **100**, 36004 (2012)
10. A. Singh, S. Ahmad, S. Puri, S. Singh, Eur. Phys. J. E **37**, 2 (2014)
11. N.J. Mottram, J.P. Newton (2014). arXiv:1409.3542
12. P.G. de Gennes, J. Prost, *The Physics of Liquid Crystals* (Oxford University, Oxford, 1995)
13. G.P. Alexander, B.G. Chen, E.A. Matsumoto, R.D. Kamien, Rev. Mod. Phys. **84**, 497 (2012)
14. N.D. Mermin, Rev. Mod. Phys. **51**, 591 (1979)
15. M. Kumar, S. Chatterjee, R. Paul, S. Puri, Phys. Rev. E **96**, 042127 (2017)
16. A. Vats, V. Banerjee, S. Puri, Europhys. Lett. **128**, 66001 (2020)
17. A. Vats, V. Banerjee, S. Puri, Soft Matter **17**, 2659 (2021)
18. G.R. Luckhurst, T.J. Sluckin, *Biaxial Nematic Liquid Crystals: Theory, Simulation and Experiment* (Wiley, 2015)
19. N.V. Priezjev, R.A. Pelcovits, Phys. Rev. E **66**, 051705 (2002)
20. A. Bhattacharjee, Inhomogeneous Phenomena in Nematic Liquid Crystals. Ph.D. thesis, 2010
21. H. Pleiner, E. Jarkova, H.W. Muler, H.R. Brand, Magnetohydrodynamics **37**, 146 (2001)
22. A. Mertelj, D. Lisjak, M. Drofenik, M. Čopič, Nature **504**, 237 (2013)
23. K. Bisht, V. Banerjee, P. Milewski, A. Majumdar, Phys. Rev. E **100**, 012703 (2019)
24. K. Bisht, Y. Wang, V. Banerjee, A. Majumdar, Phys. Rev. E **101**, 022706 (2020)
25. A. Mertelj, D. Lisjak, Liq. Cryst. Rev. **5**, 1 (2017)
26. A. Hubert, R. Schäfer, *Magnetic Domains: The Analysis of Magnetic Microstructures* (Springer Science & Business Media, 2008)
27. S. Puri, Y. Oono, J. Phys. A **21**, 755 (1988)
28. R. Paul, S. Puri, H. Rieger, Europhys. Lett. **68**, 881 (2004)
29. R. Paul, S. Puri, H. Rieger, Phys. Rev. E **71**, 061109 (2005)
30. A.J. Bray, S. Puri, Phys. Rev. Lett. **67**, 2670 (1991)
31. H. Toyoki, Phys. Rev. B **45**, 1965 (1992)
32. I.S. Gradshteyn, I.M. Ryzhik, *Table of Integrals, Series, and Products* (Academic, 2014)
33. G. Porod, O. Glatter, O. Kratky, ed. by O. Glatter, O. Kratky. (Academic, London, 17, 1982)
34. Y. Oono, S. Puri, Mod. Phys. Lett. B **2**, 861 (1988)
35. S.M. Allen, J.W. Cahn, Acta Metall. **27**, 1085 (1979)
36. I.M. Lifshitz, V.V. Slyozov, J. Phys. Chem. Solids **19**, 35 (1961)
37. E.W. Cheney, D.R. Kincaid, *Numerical Mathematics and Computing* (Cengage Learning, 2012)
38. Y. Oono, S. Puri, Phys. Rev. Lett. **58**, 836 (1987)
39. Y. Oono, S. Puri, Phys. Rev. A **38**, 434 (1988)
40. S. Puri, Y. Oono, Phys. Rev. A **38**, 1542 (1988)
41. C. Tsakonas, A.J. Davidson, C.V. Brown, N.J. Mottram, Appl. Phys. Lett. **90**, 111913 (2007)
42. N. Kumar, R. Zhang, J.J. de Pablo, M.L. Gardel, Sci. Adv. **4**, 7779 (2018)
43. F. Brochard, P.G. de Gennes, J. Phys. **31**, 691 (1970)

Chapter 4
Emergence of Biaxial Order in Ferronematics

Abstract The *biaxial phase* in nematic liquid crystals has been experimentally elusive for several decades, following its initial prediction in 1970. A notable recent experimental success came from Liu et al. [Proc. Nat. Acad. Sci. **113**, 10479 (2016)], who observed biaxiality in FNs. By leveraging the distinct spatial scales of dipolar and magneto-nematic interactions, they achieved an equilibrium configuration where the nematic phase has biaxial characteristics. The chapter provides a theoretical framework to understand the emergence of biaxial order in FNs. The quantitative estimates of biaxiality as a function of coupling strength are well supported by an analytical expression obtained by using fixed-point analysis.

4.1 Introduction

In NLCs, nematogens generally align along a primary direction **n** which gives rise to uniaxial order. In 1970, Freiser predicted the existence of a secondary direction **k** (perpendicular to **n**) based on volume exclusion calculations, leading to the concept of biaxial nematic liquid crystals [1]. Since this prediction, BNLCs have been extensively studied both experimentally and theoretically [2–8]. They are believed to provide better response times and superior display performance compared to uniaxial LCs [7, 8].

The underlying mechanism in LC applications is the Fréedericksz transition, where the molecular orientation of nematogens shifts from an ordered to a disordered state, altering their light transmission properties [7–10]. For BNLCs, it was hypothesized that such transitions could occur along more than one axis. However, the experimental identification of thermotropic BNLCs remained elusive until 2004, when three separate research groups independently confirmed the existence of the biaxial phase [11–13]. Despite this achievement, claims of an equilibrium biaxial phase were questioned in later studies [7, 8]. A detailed analysis of the current understanding of biaxiality in LCs is provided in [14]. Experimental observations on biaxial nematics revealed that the Fréedericksz transition relative to the secondary director is energetically more favorable, allowing significantly faster switching (1 ms) compared to uniaxial NLCs (15 ms) [7–10, 15]. However, stabilizing the biaxial phase remains challenging, as molecular alignment along the secondary axis is easily perturbed by thermal fluctuations [7, 8].

A recent breakthrough has come from the experiments of Liu et al., who succeeded in inducing the biaxial phase by suspending magnetic nanoparticles in an NLC medium [16]. These experiments used the distinct length scales associated with dipolar and magneto-nematic interactions to establish an equilibrium state where the magnetic moment of the nanoparticles forms an angle with the nematic director **n**. This interaction created an additional ordering direction **k** in the perpendicular plane without extra energy expenditure; see Fig. 4.1. The biaxial order was subsequently confirmed by Liu et al. from absorption spectrum and magnetic hysteresis measurements [16]. This development opens new possibilities for NLC applications that require further theoretical support. In this chapter, a theoretical framework that captures the origin of biaxiality in FNs is presented. This framework provides guidance for experimental research by offering quantitative evaluations of biaxiality based on coupling strength.

The organization of this chapter is as follows. Section 4.2 introduces the free energy and dynamical equations that govern FNs. Section 4.3 discusses the findings related to ordering kinetics and the evolution of biaxiality in FNs. Finally, Sect. 4.3 provides a summary and discussion of the results.

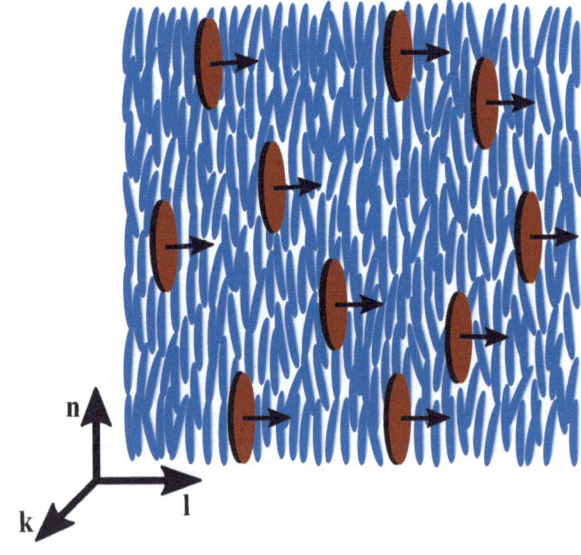

Fig. 4.1 Schematic depicting the orientations of nematic (blue) and magnetic (red) particles for the coupling limit $\gamma < 0$. Reprinted with permission from [24]. © 2022, American Physical Society. All rights reserved

4.2 Theoretical Framework

4.2.1 Coarse-Grained Free Energy for Ferronematics

Our starting point is the free energy for the FNs introduced in Sect. 3.2 constructed using the order parameter fields $\mathbf{Q}(\mathbf{r})$ and $\mathbf{M}(\mathbf{r})$:

$$G[\mathbf{Q}, \mathbf{M}] = \int d\mathbf{r} \left\{ \frac{A}{2}\text{Tr}(\mathbf{Q}^2) + \frac{C}{3}\text{Tr}(\mathbf{Q}^3) + \frac{B}{4}[\text{Tr}(\mathbf{Q}^2)]^2 + \frac{L}{2}|\nabla \mathbf{Q}|^2 \right.$$
$$\left. + \frac{\alpha^M}{2}|\mathbf{M}|^2 + \frac{\beta^M}{4}|\mathbf{M}|^4 + \frac{\kappa^M}{2}|\nabla \mathbf{M}|^2 - \frac{\gamma \mu_0}{2} \sum_{i,j=1}^{3} Q_{ij} M_i M_j \right\}. \quad (4.1)$$

Previous studies have modeled the impact of dopant particles in LCs, focusing on the coupling between ferroelectric particles' dipole moments and NLCs at the molecular level [17–20]. These models show how impurity-induced fields can behave as aligning fields, thereby enhancing the orientational order within the NLCs. Similarly, the last term in Eq. (4.1) defines a phenomenological magneto-nematic coupling, represented as a dyadic product of $\mathbf{Q}$ and $\mathbf{M}$, where γ quantifies the coupling strength. It is influenced by factors such as the dimensions, shape, and interactions of MNPs with the NLCs. This cubic coupling term is sufficient to create magneto-nematic ordering [21], and establishes the distinct orientations of magnetic and nematic components which are essential for biaxial ordering in the system [16, 22]. In more detailed studies, additional terms involving dipolar and quadrupolar interactions could improve the free energy description, particularly for examining phase transitions and critical behaviors. However, as noted in [23], such refinements are often unnecessary for dilute FN suspensions.

Chapter 3 examined the scenario where $\gamma > 0$, leading to a preference for the alignment of $\mathbf{n}$ parallel to $\mathbf{M}$. Under these conditions, LCs solely displays uniaxial ordering. Liu et al. provided experimental evidence that biaxial ordering occurs only when there is a tilt between $\mathbf{n}$ and $\mathbf{M}$. Through careful surface functionalization, they achieved a tilt angle of up to 90°. However, their optical absorbance measurements to detect the biaxial phase were limited to a tilt range of 10°–65°. Inspired by these findings, a case with $\gamma < 0$ is selected, corresponding to a scenario where the tilt angle reaches 90°. Although it is theoretically feasible to adjust the coupling term in Eq. (4.1) to allow $\mathbf{n}$ and $\mathbf{M}$ to orient at any angle, this would significantly complicate the formulation. The onset of biaxiality (indicating the presence of two distinct directions) in NLCs for negative values of γ can be illustrated through the schematic in Fig. 4.1. If $\mathbf{M}$ is oriented along the positive x-axis, the LC molecules can arrange themselves in two perpendicular directions, specifically along the y-axis and z-axis. This behavior will be demonstrated shortly, even for minimal magneto-nematic coupling values.

Several observations regarding the free energy of FNs warrant attention: (i) The lowest energy state for the nematic system when considering terms up to order

$[\text{Tr}(\mathbf{Q}^2)]^2$, is consistently uniaxial. To achieve biaxial ordering within a purely nematic framework, it becomes essential to incorporate higher-order terms, such as $[\text{Tr}(\mathbf{Q}^2)]^3$ [25, 26]. (ii) Liu et al. introduced a Frank free-energy model to characterize FNs, which focuses solely on the elastic contributions to free energy. This simplified approach fails to account for the observed biaxiality in a theoretical context. In contrast, the Landau-de Gennes free energy framework is broader and incorporates not only the Landau free energy but also the elastic energies. These additional components are crucial for determining the preferred state of LCs, be they uniaxial, biaxial, or isotropic [27, 28]. Moreover, the quantitative assessment of biaxial order, denoted as $\mathcal{T}$, can be directly derived from the $\mathbf{Q}$-tensor.

4.2.2 Time-Dependent Ginzburg–Landau Equations

The dissipative dynamics of the FNs has been modeled using the coupled TDGL equations (see Sect. 2.1) to obtain the free energy minimum. A dimensionless form of the TDGL equations is obtained by introducing the rescaled variables $\mathbf{Q} = a\mathbf{Q}'$, $\mathbf{M} = b\mathbf{M}'$, $\mathbf{r} = \zeta \mathbf{r}'$, $t = \eta t'$. The appropriate values of the scale factors are: $a = \sqrt{|A|/2B}$, $b = \sqrt{|\alpha^M|/\beta^M}$, $\zeta = \sqrt{\kappa^M/|\alpha^M|}$, $\eta = \Gamma_M^{-1}\sqrt{2B/A}$. Dropping the primes, the dimensionless evolution equations are:

$$\frac{1}{\Gamma}\frac{\partial q_1}{\partial t} = \xi_1\left[\pm 3q_1 - q^2 3q_1 + \bar{C}(6q_1^2 - 2q_2^2 - 2q_3^2 + q_4^2 + q_5^2)\right.$$
$$\left. + l\nabla^2 q_1\right] + c_0(-M_1^2 - M_2^2 + 2M_3^2), \tag{4.2}$$

$$\frac{1}{\Gamma}\frac{\partial q_2}{\partial t} = \xi_1\left[\pm q_2 - q^2 q_2 + \bar{C}(4q_1 q_2 + q_4^2 - q_5^2) + l\nabla^2 q_2\right]$$
$$+ c_0(M_1^2 - M_2^2), \tag{4.3}$$

$$\frac{1}{\Gamma}\frac{\partial q_3}{\partial t} = \xi_1\left[\pm q_3 - q^2 q_3 + \bar{C}(-4q_1 q_3 + 2q_4 q_5) + l\nabla^2 q_3\right]$$
$$+ 2c_0 M_1 M_2, \tag{4.4}$$

$$\frac{1}{\Gamma}\frac{\partial q_4}{\partial t} = \xi_1\left[\pm q_4 - q^2 q_4 + \bar{C}(2q_1 q_4 + 2q_2 q_4 + 2q_3 q_5)\right.$$
$$\left. + l\nabla^2 q_4\right] + 2c_0 M_1 M_3, \tag{4.5}$$

$$\frac{1}{\Gamma}\frac{\partial q_5}{\partial t} = \xi_1\left[\pm q_5 - q^2 q_5 + \bar{C}(2q_1 q_5 - 2q_2 q_5 + 2q_3 q_4)\right.$$
$$\left. + l\nabla^2 q_5\right] + 2c_0 M_2 M_3, \tag{4.6}$$

$$\frac{\partial M_1}{\partial t} = \xi_2\left[\pm M_1 - |\mathbf{M}|^2 M_1 + \nabla^2 M_1\right] + c_0[(q_2 - q_1)M_1$$
$$+ q_3 M_2 + q_4 M_3], \tag{4.7}$$

4.3 Results

$$\frac{\partial M_2}{\partial t} = \xi_2 \left[\pm M_2 - |\mathbf{M}|^2 M_2 + \nabla^2 M_2 \right] + c_0 [-(q_1 + q_2) M_2$$
$$+ q_3 M_1 + q_5 M_3], \qquad (4.8)$$

$$\frac{\partial M_3}{\partial t} = \xi_2 \left[\pm M_3 - |\mathbf{M}|^2 M_3 + \nabla^2 M_3 \right] + c_0 [2 q_1 M_3$$
$$+ q_4 M_1 + q_5 M_2]. \qquad (4.9)$$

Here,

$$\xi_1 = \frac{2A\beta^M}{\alpha^M} \sqrt{\frac{A}{2B}}, \quad \xi_2 = \alpha^M \sqrt{\frac{2B}{A}}, \quad \bar{C} = \frac{C}{2\sqrt{2AB}},$$

$$l = \frac{L\alpha^M}{2A\kappa^M}, \quad c_0 = \frac{\gamma \mu_0}{2}, \quad \Gamma = \frac{\alpha^M \Gamma_Q}{\beta^M \Gamma_M} \sqrt{\frac{2B}{A}},$$

$$q^2 = 3q_1^2 + q_2^2 + q_3^2 + q_4^2 + q_5^2. \qquad (4.10)$$

The $\pm$ sign denotes whether the quench temperature is below ($+$) or above ($-$) the critical temperature. Here, the case with $T < T_c^N, T_c^M$ is considered. Thus, Eqs. (4.2)–(4.9) are taken with $+$ sign, that is, both $\mathbf{Q}$ and $\mathbf{M}$ prefer the ordered state in the absence of coupling ($c_0 = 0$). The parameters ξ_1 and ξ_2 depend on the magnitudes of $\mathbf{Q}$ and $\mathbf{M}$, l is proportional to the relative elastic constant, and $\bar{C}$ determines the order of the transition. The parameter c_0 is the magneto-nematic coupling strength, and Γ determines the relative time-scales for $\mathbf{Q}$ and $\mathbf{M}$ during the evolution process. Equation (4.10) provides the values of these rescaled parameters in terms of the Landau coefficients. Note that different combinations of these coefficients can lead to the same values of the rescaled parameters. For simplicity, $\Gamma = 1, \bar{C} = 1$, and $l = 1$ are taken. Unless otherwise specified, the results are presented for $\xi_1 = \xi_2 = 1$. Note that Eqs. (4.2)–(4.9) are qualitatively similar to the corresponding equations in Sect. 3.2.2, although with different scaling. This novel approach facilitates the incorporation of magneto-nematic interaction effects into a unified coupling parameter, c_0. Consequently, the analytical solutions derived for $\mathcal{T}$ become more straightforward.

4.3 Results

The dynamical Eqs. (4.2)–(4.9) are solved using the Euler discretization method (see Sect. 2.3). The initial conditions are taken as small random fluctuations about 0, corresponding to the high-temperature disordered state for both fields. The discretization mesh sizes $\Delta x = 1$ and $\Delta t = 10^{-4}$ are used in the simulation. Periodic boundary conditions were used to simulate bulk behavior and remove edge effects. The results presented here are for system size N^3 ($N = 64$), averaged over 10 independent runs denoted by $\langle \cdots \rangle$. The evolution of Eqs. (4.2)–(4.9) provides $\{Q_{ij}\}$ and M_i at all lattice points. The physically relevant quantities $\mathbf{n}$, $\mathbf{k}$, $\mathcal{S}$ and $\mathcal{T}$ can be obtained from $\mathbf{Q}$, see the text following Eq. (2.1).

4.3.1 Ordering Kinetics

Figure 4.2 depicts the evolution of nematic morphologies (**n**) at time $t = 50$ under identical initial random configurations. In the upper panel, the snapshots show $(a) c_0 = 0$ and $(b) c_0 = -5$. Due to the inversion symmetry of the **n**-field, each point in the cubic grid is color-coded according to the key, representing one of four orientations. Domain growth occurs more rapidly in the uncoupled system compared to the FN case. The negative magneto-nematic coupling parameter γ enforces a perpendicular orientation of **n** relative to **M**. The lower panel illustrates the **n**-field at $t = 50$ for $(c) c_0 = 0$ and $(d) c_0 = -5$. In these snapshots, areas where $\mathbf{n} \perp \mathbf{M}$ are highlighted by regions where the dot product $|\mathbf{n} \cdot \mathbf{M}| < 0.05$. In configuration (c), while both **n** and **M** align, there is no restriction on their relative orientation. On the other hand, in (d) the magneto-nematic coupling enforces $\mathbf{n} \perp \mathbf{M}$.

4.3.2 Emergence of Biaxiality

Note that Chap. 3 presents extensive coarsening studies in FNs for $\gamma > 0$ which favors $\mathbf{n} \parallel \mathbf{M}$. There, a different rescaling was employed for the TDGL equations.

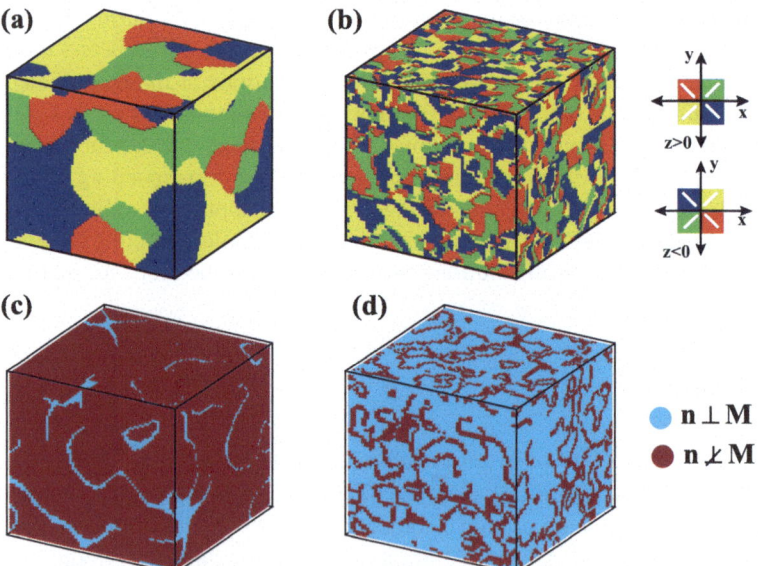

Fig. 4.2 Nematic morphologies for the cases **a** $c_0 = 0$, and **b** $c_0 = -5$ at $t = 50$. The regions are colored according to the direction of **n**, as shown in the key. The snapshots in **c** $c_0 = 0$, and **d** $c_0 = -5$ depict the regions corresponding to $\mathbf{n} \perp \mathbf{M}$ and $\mathbf{n} \not\perp \mathbf{M}$. Reprinted with permission from [24]. © 2022, American Physical Society. All rights reserved

4.3 Results

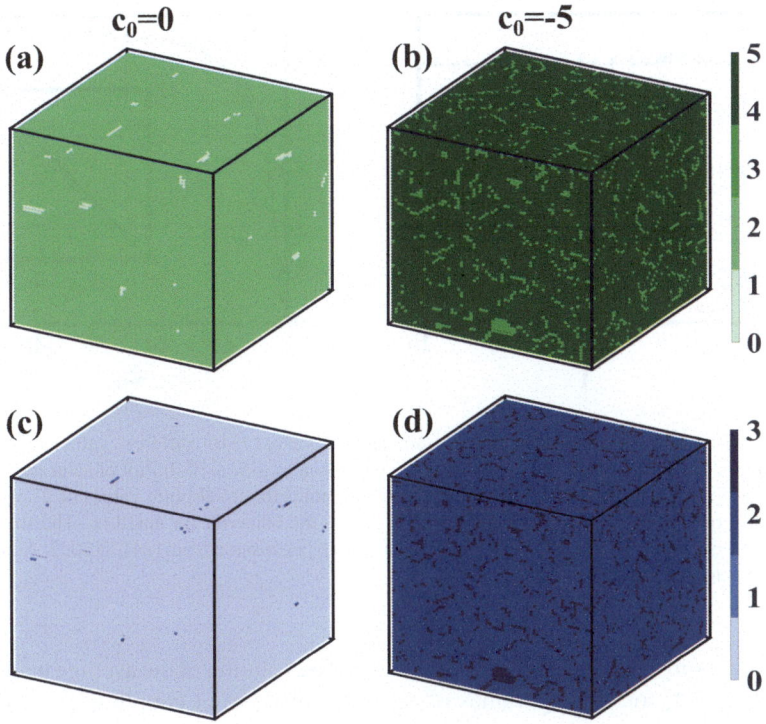

Fig. 4.3 Morphologies of the $\mathcal{S}$-field for **a** $c_0 = 0$, and **b** $c_0 = -5.0$ at $t = 50$. The regions are colored according to the magnitude of $\mathcal{S}$. The corresponding $\mathcal{T}$-field is shown below in **c** $c_0 = 0$, and **d** $c_0 = -5.0$. Reprinted with permission from [24]. © 2022, American Physical Society. All rights reserved

In that study there were two important observations: (i) slaved coarsening of fields **n** and **M** for quench temperatures intermediate to the critical temperatures of the uncoupled components, and (ii) formation of subdomain morphology separated by interfacial defects in the field **M**. Both of these features are also observed for $\gamma < 0$ for different ranges of the parameters ξ_1 and ξ_2.

To illustrate how the **Q**-field in Fig. 4.2 acquires biaxiality with the introduction of coupling, note that uniaxial LCs typically exhibit orientational order along a single primary director, **n**. When an additional order emerges in the plane perpendicular to **n**, it introduces a secondary director, **k**, indicating biaxiality within the system [27]. Figure 4.3a, b depict the order parameter $\mathcal{S}$ of the **n**-field at $t = 50$ for $c_0 = 0$ and $c_0 = -5$, where darker areas highlight regions with increased $\mathcal{S}$. It is evident that the **n**-field exhibits considerable ordering in both cases. In Fig. 4.3c, d, the corresponding order parameter $\mathcal{T}$ of the **k**-field is shown, where substantial order appears only when magneto-nematic coupling is present.

Subsequently, the average biaxiality parameter, $\langle \mathcal{T} \rangle$, is estimated by first calculating the spatial average of $\mathcal{T}(\mathbf{r}, t)$ for each run, followed by averaging over multiple

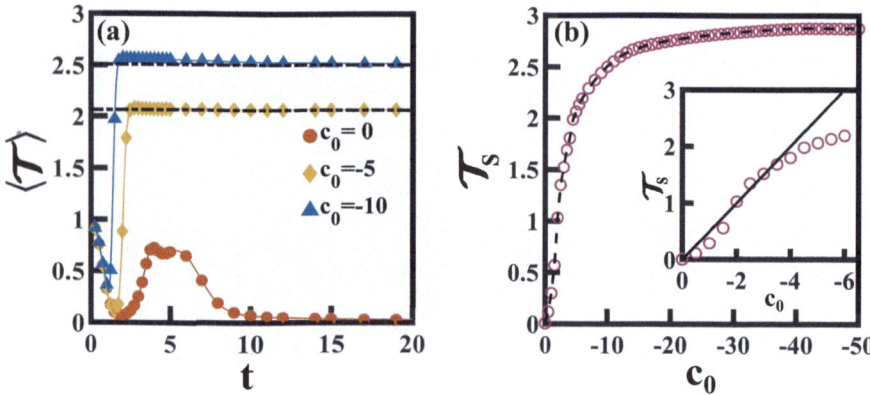

Fig. 4.4 a Plot of the average biaxiality parameter, $\langle \mathcal{T} \rangle$ versus t, for different values of c_0. The dashed lines correspond to the fixed-point values of $\mathcal{T}$ for $c_0 = -5, -10$. **b** Plot of saturation value of biaxiality parameter $\mathcal{T}_s$ versus c_0. The dashed line denotes the fixed-point values of $\mathcal{T}$, obtained numerically from the TDGL equations. The inset shows the behavior for small c_0. The solid line denotes the result in Eq. (4.23) with $\bar{C} = 1$. Reprinted with permission from [24]. © 2022, American Physical Society. All rights reserved

independent runs. Figure 4.4a depicts the temporal evolution of the average biaxiality parameter, $\langle \mathcal{T} \rangle$, for varying values of c_0. In the absence of coupling, i.e., $c_0 = 0$, $\langle \mathcal{T} \rangle$ approximately stabilizes at zero after initial transients, indicating a return to the uniaxial state. However, when $c_0 \neq 0$, $\langle \mathcal{T} \rangle$ increases over time and eventually reaches a stable saturation value $\mathcal{T}_s$ at late times. This trend is observed even for minimal values of c_0 (e.g., $c_0 = -0.05$).

These saturation values are calculated using the fixed-point solutions $\mathbf{Q}^*$ and $\mathbf{M}^*$ derived from the TDGL equations. Setting $\partial/\partial t = 0$ and $\nabla^2 = 0$ in Eqs. (4.2)–(4.9), the equations are reduced to a steady-state form, which is then solved using the Newton–Raphson method [29]. The relevant equations are:

$$\xi_1 \left[\pm 3q_1 - q^2 3q_1 + \bar{C}(6q_1^2 - 2q_2^2 - 2q_3^2 + q_4^2 + q_5^2) \right]$$
$$+ c_0(-M_1^2 - M_2^2 + 2M_3^2) = 0, \qquad (4.11)$$

$$\xi_1 \left[\pm q_2 - q^2 q_2 + \bar{C}(4q_1 q_2 + q_4^2 - q_5^2) \right] + c_0(M_1^2 - M_2^2) = 0, \qquad (4.12)$$

$$\xi_1 \left[\pm q_3 - q^2 q_3 + \bar{C}(-4q_1 q_3 + 2q_4 q_5) \right] + 2c_0 M_1 M_2 = 0, \qquad (4.13)$$

$$\xi_1 \left[\pm q_4 - q^2 q_4 + \bar{C}(2q_1 q_4 + 2q_2 q_4 + 2q_3 q_5) \right] + 2c_0 M_1 M_3 = 0, \qquad (4.14)$$

$$\xi_1 \left[\pm q_5 - q^2 q_5 + \bar{C}(2q_1 q_5 - 2q_2 q_5 + 2q_3 q_4) \right] + 2c_0 M_2 M_3 = 0, \qquad (4.15)$$

$$\xi_2 \left[\pm M_1 - |\mathbf{M}|^2 M_1 \right] + c_0 [(q_2 - q_1)M_1 + q_3 M_2 + q_4 M_3] = 0, \qquad (4.16)$$

$$\xi_2 \left[\pm M_2 - |\mathbf{M}|^2 M_2 \right] + c_0 [-(q_1 + q_2)M_2 + q_3 M_1 + q_5 M_3] = 0, \qquad (4.17)$$

$$\xi_2 \left[\pm M_3 - |\mathbf{M}|^2 M_3 \right] + c_0 [2q_1 M_3 + q_4 M_1 + q_5 M_2] = 0. \qquad (4.18)$$

4.3 Results

In Fig. 4.4a, the dashed horizontal lines indicate the numerically computed fixed-point values derived from Eqs. (4.11)–(4.18). Following this, the relationship between $\mathcal{T}_s$ and the magneto-nematic coupling strength c_0 is examined. Figure 4.4b illustrates how $\mathcal{T}_s$ varies with c_0: initially, $\mathcal{T}_s$ increases for small values of c_0 but eventually saturates as c_0 becomes larger.

Solving the coupled nonlinear Eqs. (4.11)–(4.18) across a range of c_0 values presents a significant challenge. To address this, an analytical expression for the small-c_0 dependence of $\mathcal{T}$ can be derived using a perturbative method. This involves expressing the fixed points as $\mathbf{Q}^* = \mathbf{Q}_0^* + \Delta\mathbf{Q}$ and $\mathbf{M}^* = \mathbf{M}_0^* + \Delta\mathbf{M}$, where ($\mathbf{Q}_0^*$, $\mathbf{M}_0^*$) represent the fixed points of the uncoupled equations (for $c_0 = 0$). Additionally, rotational invariance allows us to simplify by choosing

$$\mathbf{Q}_0^* = \begin{pmatrix} -q_1^* & 0 & 0 \\ 0 & -q_1^* & 0 \\ 0 & 0 & 2q_1^* \end{pmatrix}, \quad \mathbf{M}_0^* = (1, 0, 0), \qquad (4.19)$$

where

$$q_1^* = \frac{2\bar{C} + \sqrt{4\bar{C}^2 + 12}}{6}. \qquad (4.20)$$

This corresponds to $\mathbf{n}_0^*$ pointing along the z-axis, and $\mathbf{M}_0^*$ pointing along the $+x$-axis. The expressions for ($\Delta\mathbf{Q}$, $\Delta\mathbf{M}$), correct to $\mathcal{O}(c_0)$, can be obtained from Eqs. (4.11)–(4.18) with $\xi_1 = \xi_2 = 1$:

$$\Delta\mathbf{Q} = \begin{pmatrix} -\dfrac{(3 + 2\bar{C})c_0}{6\bar{C}q_1^*(1 + \bar{C}q_1^*)} & 0 & 0 \\ 0 & \dfrac{(4\bar{C}q_1^* + 3)c_0}{6\bar{C}q_1^*(1 + \bar{C}q_1^*)} & 0 \\ 0 & 0 & -\dfrac{c_0}{3(1 + \bar{C}q_1^*)} \end{pmatrix},$$

$$\Delta\mathbf{M} = \left(-\frac{c_0 q_1^*}{2}, 0, 0\right). \qquad (4.21)$$

From the $\mathbf{Q}$-tensor, the small-c_0 dependence of $\mathcal{S}$ and $\mathcal{T}$ can be obtained as

$$\mathcal{S} = \frac{(6 + 4\bar{C}^2)q_1^* + 2\bar{C} - c_0}{3(1 + \bar{C}q_1^*)} + \mathcal{O}(c_0^2), \qquad (4.22)$$

$$\mathcal{T} = -\frac{3(1 + \bar{C}q_1^*)c_0}{\bar{C}q_1^*(6q_1^* + 4\bar{C}^2 q_1^* + 2\bar{C})} + \mathcal{O}(c_0^2). \qquad (4.23)$$

The solid line in the inset of Fig. 4.4b denotes $\mathcal{T}$ versus c_0 from Eq. (4.23) with $\bar{C} = 1$. These are in very good agreement with the numerical results up to $c_0 \simeq -4.0$.

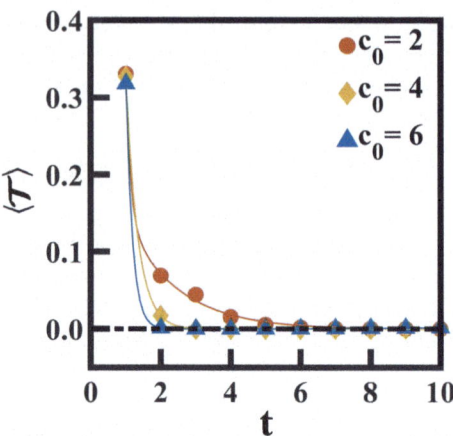

Fig. 4.5 Plot of average biaxiality parameter, $\langle \mathcal{T} \rangle$ versus t, for different values of positive c_0. Reprinted with permission from [24]. © 2022, American Physical Society. All rights reserved

For $\gamma > 0$, the resulting morphologies remain uniaxial. In Fig. 4.5, the average biaxiality parameter $\langle \mathcal{T} \rangle$ is shown for $c_0 > 0$, based on evaluations from Eqs. (4.2)–(4.9). The dashed line represents the fixed-point values of $\mathcal{T}$, calculated numerically from Eqs. (4.11)–(4.18). In particular, $\langle \mathcal{T} \rangle$ approaches these fixed point solutions at later times, confirming uniaxial order when c_0 is positive. For this scenario, straightforward analytical expressions such as Eqs. (4.22) and (4.23) are not applicable as the inclusion of higher-order terms in c_0 is necessary.

4.4 Summary and Discussion

In conclusion, this chapter has developed a framework to explain how magneto-nematic coupling induces biaxiality in FNs. The latter has gained momentum following the findings of Liu et al. [16], who experimentally observed biaxial order in FNs, a development that will advance applications in LC technology. The unique aspect of this approach lies in the inclusion of a coupling parameter $c_0 < 0$, which coerces $\mathbf{n} \perp \mathbf{M}$. Such an alignment is essential for biaxiality and is consistent with the observations of Liu et al. The framework proposed here allows for a quantitative analysis of biaxiality and its relationship with magneto-nematic coupling strength. This quantification is anticipated to enable more structured experiments and theoretical explorations.

Note that earlier experiments by Mertelj et al. [30] focused on FNs where the equilibrium state aligned $\mathbf{n}$ parallel to $\mathbf{M}$. Their findings served as the basis for later work by Liu et al. When the theoretical model is adjusted with $c_0 > 0$, it replicates the observations of Mertelj et al., promoting alignment between the nematic and magnetic order parameters [31, 32], though this configuration yields uniaxial behavior only. Consequently, adjusting model parameters within this formulation has enabled a controlled exploration of both morphologies and biaxiality.

The monograph has thus far focused on passive inclusions in NLCs, particularly FNs, exploring phase transition kinetics, morphological features, defect structures, growth patterns, and structure factors. The upcoming Chap. 5 will transition to the dynamics of NLCs with active inclusions, or living liquid crystals. Chapter 5 will introduce a new approach to studying LLC dynamics, present parameters for novel steady states and phases, and examine boundary conditions that impact LLC steady states with potential technological applications.

References

1. M.J. Freiser, Phys. Rev. Lett. **24**, 1041 (1970)
2. R. Alben, Phys. Rev. Lett. **30**, 778 (1973)
3. R. Alben, J. Chem. Phys. **59**, 4299 (1973)
4. J.P. Straley, Phys. Rev. A **10**, 1881 (1974)
5. Y. Galerne, Mol. Cryst. Liq. Cryst. **323**, 211 (1998)
6. D.W. Bruce, Chem. Rec. **4**, 10 (2004)
7. C. Tschierske, D.J. Photinos, J. Mater. Chem. **20**, 4263 (2010)
8. G.R. Luckhurst, T.J. Sluckin, *Biaxial Nematic Liquid Crystals: Theory, Simulation and Experiment* (Wiley, 2015)
9. R. Berardi, L. Muccioli, C. Zannoni, J. Chem. Phys. **128**, 024905 (2008)
10. C. Meyer, P. Davidson, D. Constantin, V. Sergan, D. Stoenescu, A. Knežević, I. Dokli, A. Lesac, I. Dozov, Phys. Rev. X **11**, 031012 (2021)
11. K. Severing, K. Saalwächter, Phys. Rev. Lett. **92**, 125501 (2004)
12. K. Merkel, A. Kocot, J.K. Vij, R. Korlacki, G.H. Mehl, T. Meyer, Phys. Rev. Lett. **93**, 237801 (2004)
13. B.R. Acharya, S. Primak, S. Kumar, Phys. Rev. Lett. **92**, 145506 (2004)
14. A. Jákli, O.D. Lavrentovich, J.V. Selinger, Rev. Mod. Phys. **90**, 045004 (2018)
15. J.H. Lee, T.K. Lim, W.T. Kim, J.I. Jin, J. App. Phys. **101**, 034105 (2007)
16. Q. Liu, P.J. Ackerman, T.C. Lubensky, I.I. Smalyukh, Proc. Nat. Acad. Sci. **113**, 10479 (2016)
17. F. Li, O. Buchnev, C. Cheon, A. Glushchenko, V. Reshetnyak, Y. Reznikov, T.J. Sluckin, J.L. West, Phys. Rev. Lett. **97**, 147801 (2006)
18. L.M. Lopatina, J.V. Selinger, Phys. Rev. Lett. **102**, 197802 (2009)
19. L.M. Lopatina, J.V. Selinger, Phys. Rev. E **84**, 041703 (2011)
20. M.V. Gorkunov, M.A. Osipov, Soft Matter **7**, 4348 (2011)
21. H. Pleiner, E. Jarkova, H.W. Muler, H.R. Brand, Magnetohydrodynamics **37**, 146 (2001)
22. A.L. Susser, S. Kralj, C. Rosenblatt, Soft Matter **17**, 9616 (2021)
23. K. Bisht, Y. Wang, V. Banerjee, A. Majumdar, Phys. Rev. E **101**, 022706 (2020)
24. A. Vats, S. Puri, V. Banerjee, Phys. Rev. E **106**, 044701 (2022)
25. A. Majumdar, Eur. J. Appl. Math. **21**, 181 (2010)
26. M.G. Forest, Q. Wang, H. Zhou, Phys. Rev. E **61**, 6655 (2000)
27. N.J. Mottram, J.P. Newton (2014). arXiv:1409.3542
28. P.G. de Gennes, J. Prost, *The Physics of Liquid Crystals* (Oxford University, Oxford, 1995)
29. E.W. Cheney, D.R. Kincaid, *Numerical Mathematics and Computing* (Cengage Learning, 2012)
30. A. Mertelj, D. Lisjak, M. Drofenik, M. Čopič, Nature **504**, 237 (2013)
31. A. Vats, V. Banerjee, S. Puri, Europhys. Lett. **128**, 66001 (2020)
32. A. Vats, V. Banerjee, S. Puri, Soft Matter **17**, 2659 (2021)

Chapter 5
Symbiotic Dynamics in Living Liquid Crystals

Abstract This chapter provides a phenomenological model for studying the dynamics of LLCs. The key ingredients of the model are (a) the Toner-Tu model for active matter, (b) the Landau-de Gennes model for nematic liquid crystals, and (c) an experimentally motivated coupling term that favors coalignment of the polarization in the active field and the nematic director. Extensive studies unfold two hitherto unobserved steady states: (i) *chimeras*, corresponding to bands of high orientational order coexisting with disorder; and (ii) *solitons*, corresponding to localized regions of order which are robust under collisions. The interplay of confinement and activity on the pattern dynamics is also studied, unraveling novel textures and tailored morphologies with potential technological applications.

5.1 Introduction

Active matter refers to assemblies of interacting particles, ranging from microscopic to macroscopic scales. These can harness energy from their surroundings and convert it into mechanical energy, resulting in self-propulsion. This category includes an extensive range of living and non-living systems, such as flocks of birds, insect swarms, herds of animals, and shoals of fish, as well as bacterial suspensions, cytoskeletal elements and protein motors, synthetic self-propelling colloids, vibrated granular media, and even human crowds [1–14]. The diversity in the constituent particles, the lack of time-reversal symmetry, and the inherent out-of-equilibrium behavior have fueled numerous experimental and theoretical studies (see [10, 14, 15] for various perspectives). The majority of these studies have focused on active matter in isotropic Newtonian fluids, though there is growing interest in active systems in non-Newtonian fluids. These complex fluids can enhance diffusivity, reduce viscosity, and exhibit directional responses that help modulate the chaotic behavior of active matter. Living liquid crystals, where active particles are introduced into NLCs, are particularly promising in this context [16–23]. NLCs themselves are clas-

---

Supplementary Information The online version contains supplementary material available at https://doi.org/10.1007/978-3-031-87799-5_5.

sic examples of anisotropic fluids with long-range or quasi-long-range order below a critical temperature, T_c^N, characterized by a preferred molecular alignment direction known as the director **n** [24, 25]. Due to this anisotropy, NLCs display distinct directional mechanical, optical, and diffusive properties [25].

In LLC studies, low concentrations of rod-shaped bacterial swimmers in lyotropic NLCs confined to quasi-2D geometries are often explored. Experiments reveal that topological defects play a vital role in the transport of bacteria through the NLC medium. For example, bacteria migrate to topological defects with charge $+1/2$ from defects with topological charge $-1/2$. Moreover, bacterial activity modifies the orientational order of the NLCs on length scales that far exceed the size of the bacteria [16, 18, 20, 26]. Such dynamics may also be relevant to other low-concentration self-propelled particle systems, including synthetic swimmers. LLCs have the potential to utilize the characteristics of active and passive matter, promising advancements in microfluidic devices for fluid transport without external pumps or pressures, synthetic systems mimicking cellular motion, and nanotechnologies for targeted drug delivery, sensing, and other biomedical applications.

A key research direction is the development of models for LLCs that can accurately mimic pattern formation when three interactions play significant roles: AM-AM, LC-LC, and AM-LC. In this scenario, the dynamics is expected to be symbiotic, involving a complex interplay between AM and LCs. This study begins with two established coarse-grained models: the Toner-Tu framework for active matter and the LdG free energy for NLCs, complemented by a coupling term inspired by experimental observation [20]. Since the LdG model excludes hydrodynamics, it lacks inherent director dynamics; this is typically incorporated using the coarse-grained TDGL equations, which are purely relaxational [27, 28]. With this theoretical basis, the impact of coupling strength on the natural states of the individual components is analyzed. A phase diagram outlining the coupling parameters predicts the emergence of novel steady states in LLCs, supported by a linear stability analysis detailed in this chapter.

In addition, boundaries are known to influence orientational order, leading to the formation of patterns such as stripes, vortices, and clusters [29–33]. Boundaries can also stabilize patterns that would otherwise be unstable in the bulk. In a significant experiment, Peng et al. [18] demonstrated the ability to induce specific bulk configurations through targeted surface treatments on bounding plates in LLC systems. The latter part of this chapter explores the use of boundary conditions to control the bulk defect configurations and guide the motion of active matter, an area of significant interest in AM physics. Confinement effects have been extensively studied for pure LCs, especially within square well geometries [34–41]. In the context of LLCs, essential questions arise: Will nematic morphologies remain stable with the addition of active particles? Can topological defects in the nematic field be manipulated to direct active particle trajectories? How do various boundary conditions shape the emergent structures from this coupling between nematic order and activity? This chapter explores these questions, uncovering intriguing and novel morphologies in confined LLC systems.

This chapter is organized as follows. In Sect. 5.2, the proposed coarse-grained modeling of LLCs is defined along with the resulting dynamical equations. The results of the numerical simulation of these equations for symbiotic dynamics in bulk LLCs are presented in Sect. 5.3. Next, Sect. 5.4 presents results from the exploration of boundary effects in LLCs. We conclude with a summary and discussion in Sect. 5.5.

5.2 Theoretical Framework

5.2.1 Model

As discussed earlier, the LdG free energy provides deep insight into NLCs [24, 42]. The TDGL equation is the appropriate framework for studying the dissipative dynamics that drives the system to its free energy minimum at the coarse-grained level [27, 28]:

$$\frac{\partial \mathbf{Q}}{\partial t} = -\Gamma_Q \frac{\delta F_Q[\mathbf{Q}]}{\delta \mathbf{Q}}. \tag{5.1}$$

Here, Γ_Q is the damping factor and sets the relaxation time scale for the nematic component. The term on the right of Eq. (5.1) is the functional derivative of the free energy functional.

The corresponding coarse-grained formulation for active matter is provided by Toner and Tu. The order parameters in the model are: (i) the local density of the active particles $\rho(\mathbf{r}, \mathbf{t})$, and (ii) the local polarization $\mathbf{P}(\mathbf{r}, t)$ that describes their average orientation [10, 14, 43–45]. Although the model was originally formulated phenomenologically using symmetry considerations, it is instructive to write the equations of motion in terms of the "free energy". The equations have already been discussed in Sect. 2.2.2, and are written as:

$$\frac{\partial \rho}{\partial t} = -v_0 \nabla \cdot (\mathbf{P}\rho) - \nabla \cdot \left(-\Gamma_\rho \nabla \frac{\delta F_a}{\delta \rho}\right), \tag{5.2}$$

$$\frac{\partial \mathbf{P}}{\partial t} = \lambda_1 (\mathbf{P} \cdot \nabla)\mathbf{P} - \Gamma_P \frac{\delta F_a}{\delta \mathbf{P}}. \tag{5.3}$$

Here, F_a is the free energy functional for the active particles (see Sect. 2.1.3 in Chap. 2). The parameter $\alpha(\rho)$ drives the order-disorder transition. For $\rho_0 < \rho_c$, the system relaxes to a disordered phase with $\mathbf{P} = 0$. On the other hand, for $\rho_0 > \rho_c$, we obtain a state of uniform orientational order with $|\mathbf{P}|^2 \sim (\rho_0/\rho_c - 1)$. Further near the transition point ($\rho_0 = \rho_c^+$), the ordered phase becomes unstable, and the system relaxes to a *banded phase* that sweeps through the system with speed v_0 [10, 45]. Additionally, there are some observations of solitons limited to the one-dimensional case [46–48].

With the coarse-grained models for both components in place, the free energy of this composite system is written as the sum of (a) free energies of the nematic and active components and (b) a suitably designed coupling term. The experiments of Genkin et al. [20] have revealed that the active particles move parallel to the nematic director. Keeping this in mind, the coupling between the nematic and active component is defined as the dyadic product of the **Q**-tensor and the polarization vector **P**. It is the lowest-order term that respects the up-down symmetry of **n** and ensures **P** ∥ **n** [49–53]. Therefore, the free energy for the LLC can be written as:

$$F[\mathbf{Q}, \rho, \mathbf{P}] = F_a + F_Q - c_0 \sum_{i,j} Q_{ij} P_i P_j, \tag{5.4}$$

where c_0 quantifies the strength of the AM-nematic interaction. Note that the coupling term can be simplified to $-(\mathbf{n} \cdot \mathbf{P})^2$, which makes it easy to see that the two components prefer co-alignment [49–53].

Substituting Eq. (5.4) into Eqs. (5.1)–(5.3) and retaining gradient terms up to the second order, the dynamical equations for LLCs are given by:

$$\frac{1}{\Gamma_Q} \frac{\partial Q_{11}}{\partial t} = \pm 2|A|Q_{11} - 4C(Q_{11}^2 + Q_{12}^2)Q_{11} + 2L\nabla^2 Q_{11} + c_0(P_1^2 - P_2^2), \tag{5.5}$$

$$\frac{1}{\Gamma_Q} \frac{\partial Q_{12}}{\partial t} = \pm 2|A|Q_{12} - 4C(Q_{11}^2 + Q_{12}^2)Q_{12} + 2L\nabla^2 Q_{12} + 2c_0 P_1 P_2, \tag{5.6}$$

$$\frac{1}{\Gamma_P} \frac{\partial P_1}{\partial t} = [-\alpha(\rho) - \beta \mathbf{P} \cdot \mathbf{P}]P_1 - \frac{v_1}{2\rho_0}\nabla_x \rho + \frac{\lambda_1}{\Gamma_P}(\mathbf{P} \cdot \nabla)P_1 + \lambda_2 \nabla_x(|\mathbf{P}|^2)$$
$$+ \lambda_3 P_1(\nabla \cdot \mathbf{P}) + \kappa \nabla^2 P_1 + 2c_0(Q_{11}P_1 + Q_{12}P_2), \tag{5.7}$$

$$\frac{1}{\Gamma_P} \frac{\partial P_2}{\partial t} = [-\alpha(\rho) - \beta \mathbf{P} \cdot \mathbf{P}]P_2 - \frac{v_1}{2\rho_0}\nabla_y \rho + \frac{\lambda_1}{\Gamma_P}(\mathbf{P} \cdot \nabla)P_2 + \lambda_2 \nabla_y(|\mathbf{P}|^2)$$
$$+ \lambda_3 P_2(\nabla \cdot \mathbf{P}) + \kappa \nabla^2 P_2 + 2c_0(Q_{12}P_1 - Q_{11}P_2), \tag{5.8}$$

$$\frac{1}{\Gamma_\rho} \frac{\partial \rho}{\partial t} = -\frac{v_0}{\Gamma_\rho}\nabla \cdot (\mathbf{P}\rho) + D_\rho \nabla^2 \rho. \tag{5.9}$$

The $\pm$ signs in Eqs. (5.5)–(5.6) refer to $T > T_c^N$ (−) and $T < T_c^N$ (+), where T_c^N is the ordering temperature of the pure nematic. Although the parameters $\lambda_2 = w/2$ and $\lambda_3 = -w$ emerge from the same term of free energy, these are treated as unrelated phenomenological parameters, as both these terms are allowed under considerations of symmetry. Note that this model does not include the hydrodynamics of the nematic. This is a reasonable assumption for situations where AM-nematic interactions are short-ranged and the nematogen's velocity is lower than the propulsion velocity of the active particle. This scenario is applicable in studies like Ref. [48] or for AM within pre-designed director patterns [18, 23].

The dimensionless form of Eqs. (5.5)–(5.9) is obtained by introducing the rescaled variables:

$$\mathbf{Q} = c_Q \mathbf{Q'}, \quad \mathbf{P} = c_P \mathbf{P'}, \quad \mathbf{r} = c_r \mathbf{r'}, \quad t = c_t t', \tag{5.10}$$

where the appropriate scale factors are

5.2 Theoretical Framework

$$c_Q = \sqrt{\frac{|A|}{2C}}; \quad c_P = \sqrt{\frac{\alpha_0}{\beta}}; \quad c_t = \frac{\beta}{\alpha_0 \Gamma_Q}\sqrt{\frac{|A|}{2C}}; \quad c_r = \sqrt{\frac{L}{|A|}}. \quad (5.11)$$

Dropping the primes, the dynamical equation for the LLCs are

$$\frac{\partial Q_{11}}{\partial t} = \xi_1 \left[\pm Q_{11} - (Q_{11}^2 + Q_{12}^2)Q_{11} + \nabla^2 Q_{11}\right] + c_0(P_1^2 - P_2^2), \quad (5.12)$$

$$\frac{\partial Q_{12}}{\partial t} = \xi_1 \left[\pm Q_{12} - (Q_{11}^2 + Q_{12}^2)Q_{12} + \nabla^2 Q_{12}\right] + 2c_0 P_1 P_2, \quad (5.13)$$

$$\frac{1}{\Gamma}\frac{\partial P_1}{\partial t} = \xi_2 \Bigg[\left(\frac{\rho}{\rho_c} - 1 - \mathbf{P}\cdot\mathbf{P}\right)P_1 - \frac{v_1'}{2\rho_0}\nabla_x \rho + \lambda_1'(\mathbf{P}\cdot\nabla)P_1 + \lambda_2'\nabla_x(|\mathbf{P}|^2)$$

$$+ \lambda_3' P_1(\nabla\cdot\mathbf{P}) + \kappa'\nabla^2 P_1\Bigg] + c_0(Q_{11}P_1 + Q_{12}P_2), \quad (5.14)$$

$$\frac{1}{\Gamma}\frac{\partial P_2}{\partial t} = \xi_2 \Bigg[\left(\frac{\rho}{\rho_c} - 1 - \mathbf{P}\cdot\mathbf{P}\right)P_2 - \frac{v_1'}{2\rho_0}\nabla_y \rho + \lambda_1'(\mathbf{P}\cdot\nabla)P_2 + \lambda_2'\nabla_y(|\mathbf{P}|^2)$$

$$+ \lambda_3' P_2(\nabla\cdot\mathbf{P}) + \kappa'\nabla^2 P_2\Bigg] + c_0(Q_{12}P_1 - Q_{11}P_2), \quad (5.15)$$

$$\frac{1}{\Gamma'}\frac{\partial \rho}{\partial t} = -v_0'\nabla\cdot(\mathbf{P}\rho) + D_\rho'\nabla^2\rho. \quad (5.16)$$

The dimensionless parameters in Eqs. (5.12)–(5.16) are:

$$\xi_1 = \frac{2|A|\beta}{\alpha_0}\sqrt{\frac{|A|}{2C}}, \quad \xi_2 = \frac{\alpha_0}{2}\sqrt{\frac{2C}{|A|}},$$

$$v_1' = \frac{v_1}{\alpha_0}\sqrt{\frac{\beta|A|}{\alpha_0 L}}, \quad v_0' = \frac{v_0}{\Gamma_\rho}\sqrt{\frac{\alpha_0|A|}{\beta L}},$$

$$\Gamma = \frac{\beta|A|\Gamma_P}{\alpha_0 \Gamma_Q C}, \quad \Gamma' = \frac{\beta \Gamma_\rho}{\alpha_0 \Gamma_Q}\sqrt{\frac{|A|}{2C}},$$

$$\kappa' = \frac{\kappa|A|}{\alpha_0 L}, \quad D_\rho' = \frac{D_\rho |A|}{L},$$

$$\lambda_1' = \frac{\lambda_1}{\Gamma_P}\sqrt{\frac{|A|}{\alpha_0 \beta L}}, \quad \lambda_2' = \lambda_2 \sqrt{\frac{|A|}{\alpha_0 \beta L}}, \quad \lambda_3' = \lambda_3 \sqrt{\frac{|A|}{\alpha_0 \beta L}}. \quad (5.17)$$

Before proceeding further, it is important to discuss the choice of parameters. The quantities ξ_1 and ξ_2 set the relative magnitudes of **Q** and **P**, and are taken as 1 in the simulations. In dimensional units, $v_0 > 0$ is the speed of the active particle and the stable state exists only if $v_1 > 0$ [46]. The corresponding rescaled parameters are assigned the values $v_0' = 0.5$, $v_1' = 0.25$. The dimensional parameters Γ_P, Γ_Q and Γ_ρ set the relaxation scales of **P**, **Q** and ρ, respectively. Γ and Γ' measure the relative time scales and are set to 1. Similarly, κ' and D_ρ' set the relative values of the elastic scales and are assigned the value 1. In dimensional terms, the linear stability analysis yields that non-trivial states arise under the conditions $\lambda_1/\Gamma_P + \lambda_2 + \lambda_3 < 0$ and $\lambda_2 = -\lambda_3$ [10, 14]. These conditions are invariant under the above scaling and $\lambda_1' = -0.5$, $\lambda_2' = -0.5$, $\lambda_3' = 0.5$ are chosen for simplicity. It is evident that the parameters selected above entail a certain degree of flexibility. However, it is crucial to emphasize that the numerical results presented here do not change significantly as the aforementioned values are altered, provided that the specified signs are maintained. In our simulations, the coupling constant c_0 will be allowed to vary.

5.2.2 Fixed-Point Solutions and Linear Stability Analysis

The dimensionless Eqs. (5.12)–(5.16) govern the evolution of the LLC to its steady state. It is useful to study the fixed point solutions (FP) ($\mathbf{Q}^*$, $\mathbf{P}^*$), as these dictate the nature of the domains and steady states formed during evolution. To determine the FP solutions for the coupled system, $\partial/\partial t$ and ∇ are set to 0 in Eqs. (5.12)–(5.9) with $\xi_1 = \xi_2 = 1$:

$$\pm Q_{11}^* - (Q_{11}^{*\,2} + Q_{12}^{*\,2})Q_{11}^* + c_0(P_1^{*2} - P_2^{*2}) = 0, \quad (5.18)$$

$$\pm Q_{12}^* - (Q_{11}^{*\,2} + Q_{12}^{*\,2})Q_{12}^* + 2c_0 P_1^* P_2^* = 0, \quad (5.19)$$

$$(g_0 - |\mathbf{P}^*|^2)P_1^* + c_0(Q_{11}^* P_1^* + Q_{12}^* P_2^*) = 0, \quad (5.20)$$

$$(g_0 - |\mathbf{P}^*|^2)P_2^* + c_0(Q_{12}^* P_1^* - Q_{11}^* P_2^*) = 0, \quad (5.21)$$

where $g_0 = \rho_0/\rho_c - 1$. The conservation law dictates that the homogeneous FP solution of Eq. (14) is $\rho = \rho_0$. A trivial solution for Eqs. (5.18)–(5.21) is $Q_{11}^* = 0$, $Q_{12}^* = 0$, $P_1^* = 0$, $P_2^* = 0$, which corresponds to a disordered state for both components.

The non-trivial FPs are rotationally invariant and can be expressed as:

$$Q_{11}^* = r_Q \cos 2\theta, \quad Q_{12}^* = r_Q \sin 2\theta; \quad P_1^* = r_P \cos \theta, \quad P_2^* = r_P \sin \theta. \quad (5.22)$$

Here, θ is the arbitrary angle between $\mathbf{P}^* \parallel \mathbf{n}^*$ and the x-axis. We choose $\theta = 0$ without any loss of generality. This choice of θ corresponds to $Q_{11}^* = r_Q$, $P_1^* = r_P$ and $Q_{12}^* = P_2^* = 0$. The substitution of these values in Eqs. (5.18)–(5.21) simplifies them to

$$-r_Q^3 + (\pm 1 + c_0^2)r_Q \pm c_0|g_0| = 0, \quad (5.23)$$

$$r_P^2 = c_0 r_Q \pm |g_0|. \quad (5.24)$$

Here, the first $\pm$ sign in Eq. (5.23) signifies $T < T_c$ (+) or $T > T_c$ (−). The $\pm$ sign with $|g_0|$ is dictated by whether $\rho_0 > \rho_c$ (+) or $\rho_0 < \rho_c$ (−). These equations are

5.2 Theoretical Framework

Table 5.1 Fixed point (FP) solutions for *Cases 1-3*. Reprinted with permission from [54]. © 2023, American Physical Society. All rights reserved

Cases	FP solutions $(Q_{11}^*, Q_{12}^*, P_1^*, P_2^*) = (r_Q, 0, r_P, 0)$
Case 1 ($T > T_c^N$, $\rho_0 = \rho_c^+$)	$r_Q = -2^{1/3}(1+c_0^2)a_1^{-1/3} + a_1^{1/3}(2^{1/3}3)^{-1}$ $r_P^2 = c_0 r_Q + \|g_0\|$ $a_1 = 27\|g_0\|c_0 + \sqrt{(27\|g_0\|c_0)^2 + 4(3-3c_0^2)^3}$
Case 2 ($T < T_c^N$, $\rho_0 = \rho_c^-$)	$r_Q = 2^{1/3}(1+c_0^2)a_1^{-1/3} + a_1^{1/3}(2^{1/3}3)^{-1}$ $r_P^2 = c_0 r_Q - \|g_0\|$ $a_1 = -27\|g_0\|c_0 + \sqrt{(27\|g_0\|c_0)^2 + 4(3-3c_0^2)^3}$
Case 3 ($T < T_c^N$, $\rho_0 = \rho_c^+$)	$r_Q = 2^{1/3}(1+c_0^2)a_1^{-1/3} + a_1^{1/3}(2^{1/3}3)^{-1}$ $r_P^2 = c_0 r_Q + \|g_0\|$ $a_1 = 27\|g_0\|c_0 + \sqrt{(27\|g_0\|c_0)^2 + 4(3-3c_0^2)^3}$

solved for arbitrary values of c_0. The FPs thus obtained are given in Table 5.1 for all cases.

Next, let us evaluate the stability of the fixed point solutions (FP) (ρ_0, $\mathbf{P}^*$, $\mathbf{Q}^*$). To understand the evolution of small fluctuations around these solutions, we replace $(\rho, \mathbf{P}, \mathbf{Q}) \equiv (\rho_0 + \Delta\rho, \mathbf{P}^* + \Delta\mathbf{P}, \mathbf{Q}^* + \Delta\mathbf{Q})$ in Eqs. (5.12)–(5.16). It proves advantageous to work with Fourier transformed fluctuations [$\Delta\rho(\mathbf{k},t)$, $\Delta\mathbf{P}(\mathbf{k},t)$, $\Delta\mathbf{Q}(\mathbf{k},t)$]. These fluctuations give rise to linearized equations that can be succinctly expressed in vector notation:

$$\frac{\partial \Phi(\mathbf{k},t)}{\partial t} = W(\mathbf{k}) \cdot \Phi(\mathbf{k},t), \quad (5.25)$$

where $\Phi(\mathbf{k},t) = [\Delta\rho(\mathbf{k},t), \Delta P_1(\mathbf{k},t), \Delta P_2(\mathbf{k},t), \Delta Q_{11}(\mathbf{k},t), \Delta Q_{12}(\mathbf{k},t)]$. The quantity $W(\mathbf{k})$ is a 5×5 matrix:

$$W = \begin{pmatrix} iv_0'(k_x P_1^* + k_y P_2^*) \\ -D_\rho'(k_x^2 + k_y^2) & ik_x v_0' \rho_0 & ik_y v_0' \rho_0 & 0 & 0 \\[1ex] \frac{P_1^*}{\rho_c} + \frac{ik_x v_1'}{2\rho_0} & \frac{\rho_0}{\rho_c} - 1 - 3P_1^{*2} - P_2^{*2} \\ -ik_x(\lambda_1' + 2\lambda_2' + \lambda_3')P_1^* \\ -ik_y \lambda_1' P_2^* - \kappa'(k_x^2 + k_y^2) \\ +c_0 Q_{11}^* & \begin{matrix}-2P_1^* P_2^* - 2ik_x \lambda_2' P_2^* \\ -ik_y \lambda_3' P_1^* + c_0 Q_{12}^*\end{matrix} & c_0 P_1^* & c_0 P_2^* \\[1ex] \frac{P_2^*}{\rho_c} + \frac{ik_y v_1'}{2\rho_0} & \begin{matrix}-2P_1^* P_2^* - 2ik_y \lambda_2' P_1^* \\ -ik_x \lambda_3' P_2^* + c_0 Q_{12}^*\end{matrix} & \begin{matrix}\frac{\rho_0}{\rho_c} - 1 - 3P_2^{*2} - P_1^{*2} \\ -ik_y(\lambda_1' + 2\lambda_2' + \lambda_3')P_2^* \\ -ik_x \lambda_1' P_1^* - \kappa'(k_x^2 + k_y^2) \\ -c_0 Q_{11}^*\end{matrix} & -c_0 P_2^* & c_0 P_1^* \\[1ex] 0 & 2c_0 P_1^* & -2c_0 P_2^* & \begin{matrix}\pm 1 - 3Q_{11}^{*2} \\ -Q_{12}^{*2} \\ -(k_x^2 + k_y^2)\end{matrix} & -2Q_{11}^* Q_{12}^* \\[1ex] 0 & 2c_0 P_2^* & 2c_0 P_1^* & -2Q_{11}^* Q_{12}^* & \begin{matrix}\pm 1 - 3Q_{12}^{*2} \\ -Q_{11}^{*2} \\ -(k_x^2 + k_y^2)\end{matrix} \end{pmatrix}$$

$$(5.26)$$

As usual, the eigenvalues $\{\bar{\lambda}_i\}$ and eigenvectors of $W(\mathbf{k})$ determine the stability of the FP. If any $\bar{\lambda}_i > 0$, the fluctuations in the corresponding eigendirection grow exponentially in time, and the FP is unstable. To examine the stability of the disordered solution, we set $P_1^* = P_2^* = Q_{11}^* = Q_{12}^* = 0$ in Eq. (5.26). It is clear that the coupling terms do not contribute at the linear level as they are quadratic in P_i and Q_{ij}. Thus, the stability properties of the trivial disordered FP are the same as those of the LC and AM separately.

For non-trivial FPs, the analysis is more complicated, even after setting $P_2^* = Q_{12}^* = 0$. The $\{\bar{\lambda}_i(\mathbf{k})\}$ are determined numerically as a function of $\mathbf{k}$ to check if any of the values are above 0. For example, consider the phase diagram in Fig. 5.1(i). For large values of $\rho_0 - \rho_c$, the system lies in the ordered state of *Case 1* in Table 5.1. Thus, all eigenvalues are negative-definite for this state. The value of $\rho_0 - \rho_c$ is reduced while keeping c_0 constant, and the point where the first instability occurs is investigated. This signals the onset of a non-trivial ordered state with spatial inhomogeneity, identified as a chimera. This method is used to obtain the dashed line in Fig. 5.1(i).

Case 2 also has a non-trivial FP where $Q_{11}^* = 1$, $Q_{12}^* = P_1^* = P_2^* = 0$, i.e., the LC is ordered and AM is disordered. The dotted line shown later in Fig. 5.3a denotes the boundary where this isotropic state becomes unstable, foreshadowing the onset of order in both fields.

5.3 Results for LLCs in Bulk

The key element of our current theoretical modeling is to understand the interplay of AM-NLC coupling in LLCs. The focus here is on understanding the effect of the coupling strength c_0 on the dynamical evolution of active and nematic fields. The three cases which provide interesting outcomes are *Case 1*: $T > T_c, \rho_0 = \rho_c^+$; *Case 2*: $T < T_c, \rho_0 = \rho_c^-$; *Case 3*: $T < T_c, \rho_0 = \rho_c^+$. Here, ρ_c^+ (ρ_c^-) corresponds to a density slightly above (below) the critical density ρ_c. $\rho_c = 0.5$ is taken without loss of generality. For each of the three cases, Eqs. (5.12)–(5.16) are solved numerically by Euler discretization with an isotropic Laplacian on an N^2 lattice ($N = 128$). Periodic boundary conditions are imposed in both directions [55] to eliminate edge effects and mimic the bulk system. The discretization mesh sizes are chosen to be $\Delta t = 0.01$ and $\Delta x = 1.0$. The initial conditions for $\mathbf{Q}$ and $\mathbf{P}$ are chosen as small fluctuations about zero, which mimics the disordered state. The corresponding initial state for ρ is small fluctuations around the mean density ρ_0. All statistical quantities have been averaged over 10 independent initial conditions unless otherwise stated.

The consequences of AM-LC coupling are first examined for the system designated as *Case 1*. When uncoupled ($c_0 = 0$), stability analysis reveals a disordered nematic state with $\mathcal{S} \approx 0$ and a banded pattern in the active component. Figure 5.1 shows how the active and nematic fields evolve with $\rho_0 = \rho_c^+ = 0.52$ for various values of c_0. The subfigures (a) and (b) illustrate the active field density at $t = 10^2$ and $t = 10^4$ for $c_0 = 0.5$, with the white arrows indicating the direction of the $\mathbf{P}$-field,

5.3 Results for LLCs in Bulk

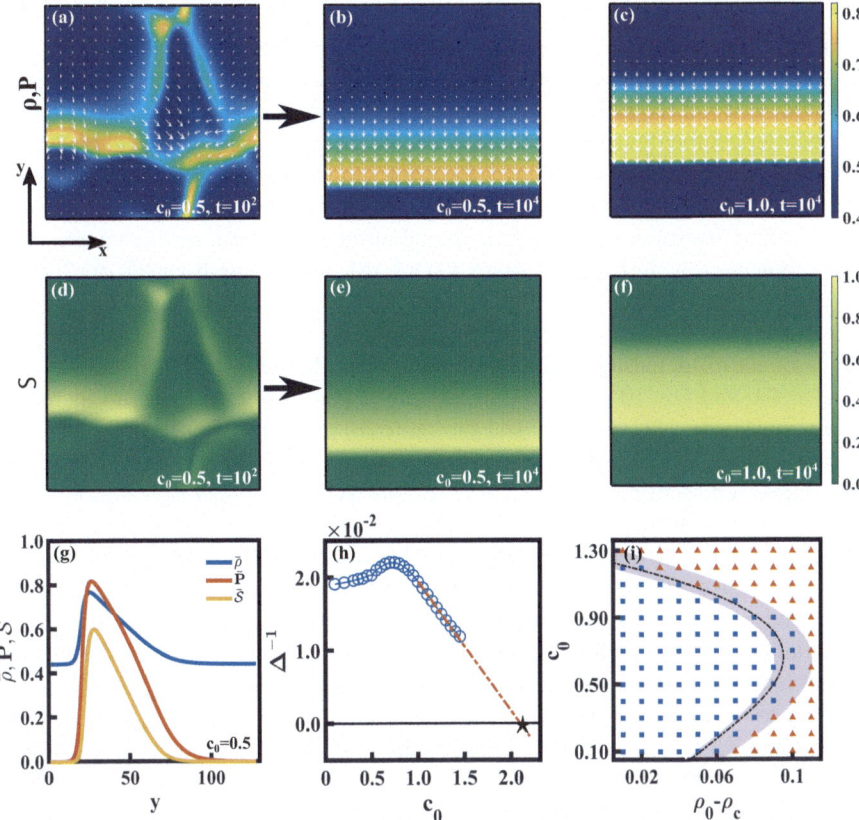

Fig. 5.1 Morphology snapshots for the active field (first row) and nematic field (second row) in *Case 1* ($T > T_c$, $\rho_0 = \rho_c^+ = 0.52$) for specified values of (t, c_0). The color bar in the top row indicates the density (ρ) of the active field; the white arrows represent the direction and magnitude of the polarization field (**P**). The color bar in the second row shows the orientational order $\mathcal{S}$ in the nematic, see text for details. Subfigure **g** shows the variation of $\bar{\rho}$, $\bar{\mathbf{P}}$ and $\bar{\mathcal{S}}$ with y for morphologies (**b**) and (**e**), where the bar indicates an average along the x-direction. Subfigure **h** shows the dependence of the inverse bandwidth Δ^{-1} on the coupling c_0. The dashed line corresponds to $\Delta^{-1} = c_0^* - c_0$, with $c_0^* = 2.1$. Subfigure **i** shows the phase diagram demarcating the ordered (▲) and chimera (■) states. The dashed line indicates the analytical phase boundary obtained in Sect. 5.2.2, while the smeared region indicates the approximate numerical counterpart. The smeared region will reduce to the analytical results for infinite system size and $\Delta x, \Delta t \to 0$. Reprinted with permission from [54]. © 2023, American Physical Society. All rights reserved

proportional in length to the vector's magnitude. The banded configuration remains evident in the active matter reflecting the uncoupled state. Within this banded phase, the ordered regions (high P) coexist with the disorder (low P), forming what is known as a *chimera* state [56, 57]. Figure 5.1a–b depict the coarsening to the chimera state, and it propagates through the system at speed v_0. The corresponding nematic field patterns appear in Figure 5.1d–e, where the color bar denotes the normalized orientational order parameter $\mathcal{S}$, with values $\mathcal{S}_m \simeq 0.67$ in (d) and $\mathcal{S}_m \simeq 0.61$ in (e). The coupling imprints the chimera state on the nematic component, which also moves at v_0. Notably, while the nematic particles remain stationary, their orientational order and disorder exhibit dynamic behavior. This unique steady state is depicted in Movie 1, provided at the end of the chapter. Figure 5.1g presents the profiles of $\bar{\rho}$, $\bar{P}$, and $\bar{\mathcal{S}}$ along y in the steady state, averaged across the x-axis. The synchronized variations in these quantities confirm their spatial alignment. These patterns correspond to traveling wave solutions of Eqs. (5.12)–(5.16), moving with speed v_0. Solving the associated differential equations numerically yields the inhomogeneous profiles shown in Fig. 5.1g.

The effect of increasing coupling strength is further analyzed in Fig. 5.1c and f for $c_0 = 1.0$ and $t = 10^4$, showing an expanded band width (Δ) and an increased orientational order ($\mathcal{S}_m \simeq 1.79$). The subfigure (h) plots Δ^{-1} as a function of c_0. At a critical point $c_0^* \simeq 2.1$, the system transitions to a uniform state with $\Delta^{-1} = 0$. The dashed line, representing $\Delta^{-1} = c_0^* - c_0$, aligns well with higher c_0 data. The slight deviation in c_0^* is attributed to finite simulation sizes. Subfigure (i) presents a phase diagram in the (c_0, ρ_0) plane, identifying stable regions for the chimera and ordered states. The phase boundary, indicated by a dashed line, is derived analytically, while the shaded region shows the numerically determined boundary, which depends on the initial conditions and the size of the system. In particular, a reentrant transition is observed for certain ρ_0 values, where the system changes from ordered to chimera and back to order as c_0 increases.

For *Case 2* ($T < T_c$, $\rho_0 = \rho_c^- = 0.48$), the uncoupled state ($c_0 = 0$) yields an ordered nematic phase with non-zero $\mathcal{S}$, while the fields ρ and **P** remain disordered. Upon coupling, significant changes occur: the active field exhibits a chimera, even with $\rho_0 = \rho_c^-$. The initially ordered nematic state also transforms into a chimera, with Movie 2 at the end of the chapter showcasing this evolution. Furthermore, elusive 2D *soliton* structures emerge for specific c_0 and ρ_c^- values, with a simulation probability of approximately 0.1. Soliton solutions have a long-standing place in the study of integrable PDEs [58–60], and while most solitonic equations (e.g., Korteweg-de Vries and nonlinear Schrödinger equations) are one-dimensional, this model of LLCs provides a rare instance of higher-dimensional solitons.

Figure 5.2 illustrates the time progression of the ρ-field (top row) and the nematic field (bottom row) for a coupling constant $c_0 = 0.1$ at $t = 800$, $t = 1000$, and $t = 1200$. In the active configurations, the white arrows indicate the polarization field within regions of high density ($\rho > 0.6$). Initially, a localized structure (denoted as L_1) moves right at $t = 800$ and then encounters a complex nonlinear interaction with other structures moving right at $t = 1000$. Despite this interaction, L_1 re-emerges with its original form intact after the collision. This preservation of shape during movement and interactions is indicative of soliton-like behavior and can be viewed in Movie 3 at the end of the chapter. In particular, the proposed LLC model operates as a dissipative rather than a Hamiltonian system, making the traditional explanation of solitons through complete integrability and infinite constants of motion inapplicable.

5.3 Results for LLCs in Bulk

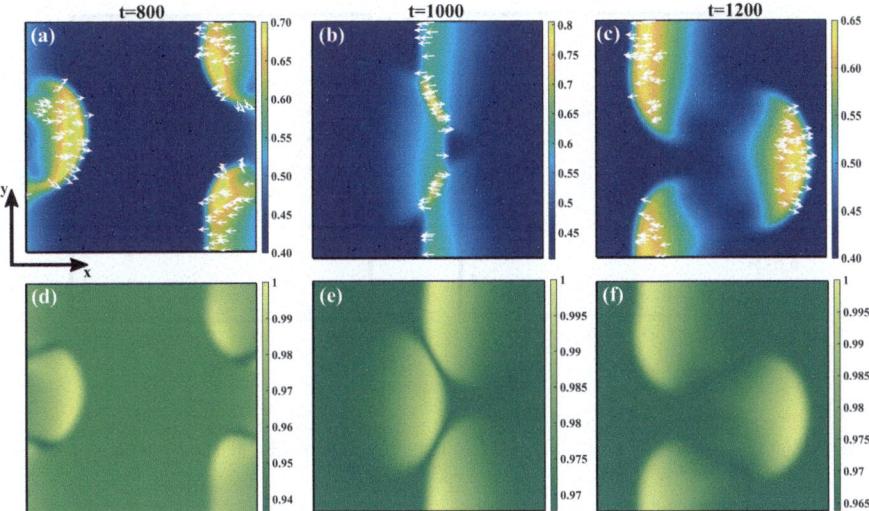

Fig. 5.2 Morphology snapshots of the active field (top row) and nematic field (bottom row) for *Case 2* with $T < T_c$, $\rho_0 = \rho_c^- = 0.48$ and $c_0 = 0.1$. The arrows in the active morphologies correspond to the polarization field in the high-density regions ($\rho > 0.6$), and denote the direction of motion of the active field. The S-field is normalized by **d** $S_m = 2.104$, **e** $S_m = 2.066$, **f** $S_m = 2.0737$ respectively. Reprinted with permission from [54]. © 2023, American Physical Society. All rights reserved

More analytical work is necessary to fully understand this stable dynamics, which is outside the scope of this study.

The phase diagrams for *Case 2* and *Case 3* are shown in Fig. 5.3a and b, respectively. For *Case 2* [Fig. 5.3a], the interaction between the nematic and active systems induces structural organization in the active field even in a nominally disordered state ($\rho_0 = \rho_c^-$). Linear stability analysis indicates the threshold for transition from a disordered to an ordered state when $c_0 + \rho/\rho_c - 1 > 0$, represented by the dotted line. Within a certain intermediate range of c_0, two distinct regions appear: one exhibits both 1D chimera states and higher-dimensional soliton states, while another supports only the chimera state. At higher c_0 values, the ordering in the nematic component propagates to the active matter, leading both fields into an ordered configuration. For *Case 3* [Fig. 5.3b], both nematic and active components are inherently ordered with $T < T_c$ and $\rho_0 = \rho_c^+$. The range supporting chimera states decreases as the value of $(\rho_0 - \rho_c)$ increases. Beyond a critical coupling threshold, $c_0 > c_0^*(\rho_0)$, the system transitions to an entirely ordered phase. In both diagrams, the analytical phase boundary from the linear stability analysis in Sect. 5.2.2 is indicated by the dashed line, while the shaded area represents the approximate phase boundary derived from numerical simulations.

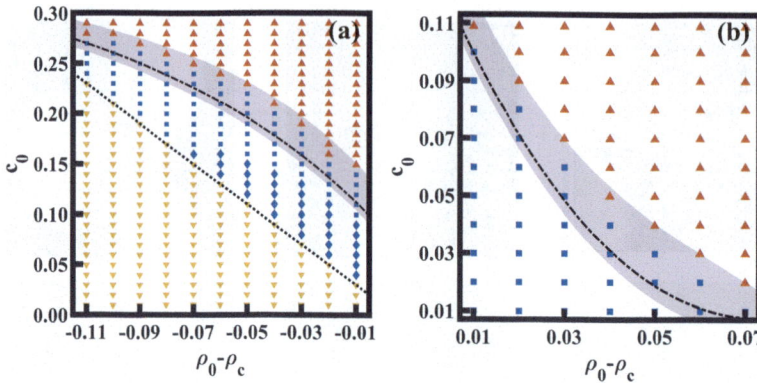

Fig. 5.3 Phase diagram for **a** *Case 2*: $T < T_c$, $\rho_0 = \rho_c^-$; and **b** *Case 3*: $T < T_c$, $\rho_0 = \rho_c^+$ showing different phases: disordered (▼), chimera (■), soliton plus chimera (♦), and ordered (▲). The phase boundaries shown by the dotted and dashed lines are obtained analytically. The smear indicates the corresponding numerical phase boundary for the chimera → ordered transition. Reprinted with permission from [54]. © 2023, American Physical Society. All rights reserved

5.4 Surface-Directed Dynamics in LLCs

The chapter thus far has explored the bulk dynamics of the symbiotic interactions within the model system. This section shifts focus to controlling the dynamics of AM and LCs by introducing surfaces that impose specific boundary conditions [61]. This approach is motivated by potential applications where control over pattern dynamics could be beneficial. To provide a foundation, it is useful to outline some quantitative aspects of the boundary conditions and corresponding solutions seen in both pure LC and AM systems. Here, the study examines various configurations, specifically combinations of planar boundary conditions (B_P) and homeotropic (B_H) for both AM and LC components, as illustrated in Fig. 5.4. Under planar conditions, the vector order parameter (**n** for LCs or **P** for AM) is aligned parallel to the walls, while homeotropic conditions anchor it perpendicular to the surfaces. Additional boundary conditions can be constructed by (a) mixing B_P and B_H for **n** and **P** on different surfaces, as seen in Fig. 5.4; and (b) reversing **P** → −**P** on one or more surfaces in the configurations shown in the lower frames of Fig. 5.4. However, this work limits its scope to the conditions presented in Fig. 5.4 for simplicity.

These boundary conditions have been extensively studied for LC square well systems [34, 36, 37, 49, 50] and are feasible for implementation in experimental settings. For instance, chemically treating the surface can create anchoring preferences for the nematic director in LCs. Alternative methods for controlling the orientation of the director include lithography, surface anchoring techniques, flow alignment, or the application of external fields [25, 62]. In the context of active matter, planar boundary conditions are most widely adopted, and geometric constraints often suffice to

5.4 Surface-Directed Dynamics in LLCs

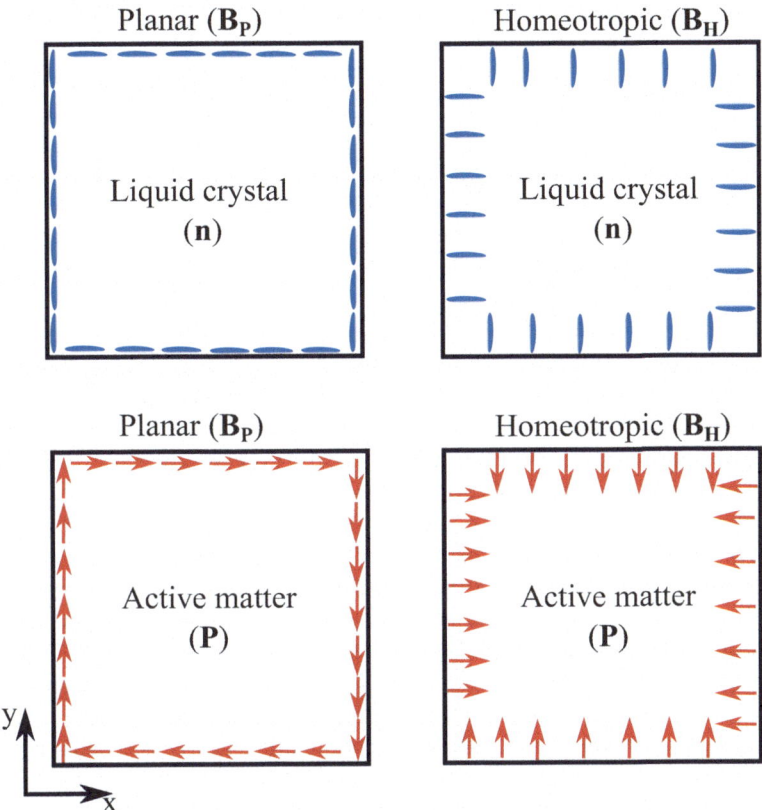

Fig. 5.4 Schematic depicting planar (B_P) and homeotropic (B_H) boundary conditions for nematic (upper frames) and active (lower frames) components. Reprinted with permission from [61]. © 2024, American Physical Society. All rights reserved

produce these effects in experimental scenarios. For homeotropic boundary conditions, studies indicate that confining walls can be engineered using different types of particles, with tunable interactions between active and wall particles to establish the preferred boundary orientations [63]. Techniques such as physical barriers, optical trapping, surface patterning, and chemical modification are also commonly applied to establish boundary conditions in active matter systems [15].

Before discussing the results, it is worth mentioning some computational details for the numerical studies. Here, Eqs. (5.12)–(5.16) are solved using boundary conditions in Fig. 5.4 for **n** and **P**. However, periodic boundary conditions are used for ρ. The parameters are such that $T < T_c$ (+ sign in Eqs. (5.12)–(5.16)) and $\rho = \rho_c^+ = 0.52$. The steady states presented hereafter remain robust with 10 different initial conditions and maintain their structure throughout our simulation time of 2×10^5.

To establish a reference point for analyzing the effects of coupling, it is useful to examine the behavior of planar and homeotropic boundary conditions applied to uncoupled systems, i.e., with $c_0 = 0$. Various studies on NLCs [34, 36–40, 49, 64] and active matter systems [29–33, 65–72] have explored these boundary conditions individually. In this uncoupled configuration, both the nematic and active fields evolve independently. Figure 5.5a, b depict the nematic morphologies at $t = 10^4$ with planar (B_P) and homeotropic (B_H) boundary conditions.

The color map represents the magnitude of the order parameter, S, and reveals topological defects with charges of $+1/2$ and $-1/2$. These defects, located at points where the director field changes by $+\pi$ or $-\pi$ on a clockwise rotation, demonstrate how the director aligns diagonally across the square to minimize bulk defect formation under strong boundary anchoring. However, partial defects appear near the corners due to a $\pi/2$ rotation in the director from the perpendicular edge alignment, as experimentally observed in LC systems confined in square wells with homeotropic boundary conditions [36]. The morphologies of the **P**-field for the active component are shown in Fig. 5.5c, d, where arrows denote the vector orientation of **P**. Under planar anchoring (B_P), a single $+1$ defect forms and circulates within the domain, a result of boundary anchoring that produces a distinct vortex pattern. The advection term propagates this vortex through the bulk, creating swirling active flows, a phenomenon reported by Sokolov et al. for circular substrates with planar boundary conditions [21]. Homeotropic anchoring (B_H), in contrast, results in multiple dynamically interacting defects. Strong anchoring with opposing directions on adjacent boundaries inhibits a single, stable defect pattern, leading instead to a chaotic state with ongoing defect generation and annihilation. Lastly, the density field ρ, shown in Fig. 5.5e, f, exhibits variations influenced by the ρ-**P** coupling in the TT equations. Notably, the nematic field configurations (Fig. 5.5a, b) achieve stability with fixed-point values, while the **P**-field in the active component (Fig. 5.5c, d) remains dynamic in nature.

The impact of coupling on surface-directed dynamics is now examined. Focusing first on the coupled system under planar boundary conditions (B_P), Fig. 5.6 shows snapshots at $t = 10^4$ of both nematic and active components for $c_0 = 0.1$ (top row) and $c_0 = 1.0$ (bottom row).

Figure 5.6a and d display the S-field (color-coded) with the director field **n** illustrated by rods. The corresponding **P**-field and its magnitude are presented in Fig. 5.6b and e, where the alignment between **n** and **P** is evident due to the coupling. Importantly, unlike the static configurations observed in Fig. 5.5a, NLC morphologies here achieve a dynamic steady state, characterized by two co-rotating topological defects with equal $+1/2$ charges, maintained at a fixed distance d_0 in the center of the square well. This nonequilibrium configuration is unique to the coupled system and cannot be realized in an isolated system at equilibrium. As c_0 increases, the defects move closer together. The **P**-field forms a central vortex of $+1$ charge, visible in Fig. 5.6b and e. The density field ρ, shown in Fig. 5.6c and f, exhibits significant variation in the highly coupled case ($c_0 = 1.0$), where dilute regions correlate with the vortex core where $|\mathbf{P}| \approx 0$. Beyond boundary-induced anchoring, the nematic alignment is further influenced by coupling with active fields through the coupling constant c_0.

5.4 Surface-Directed Dynamics in LLCs

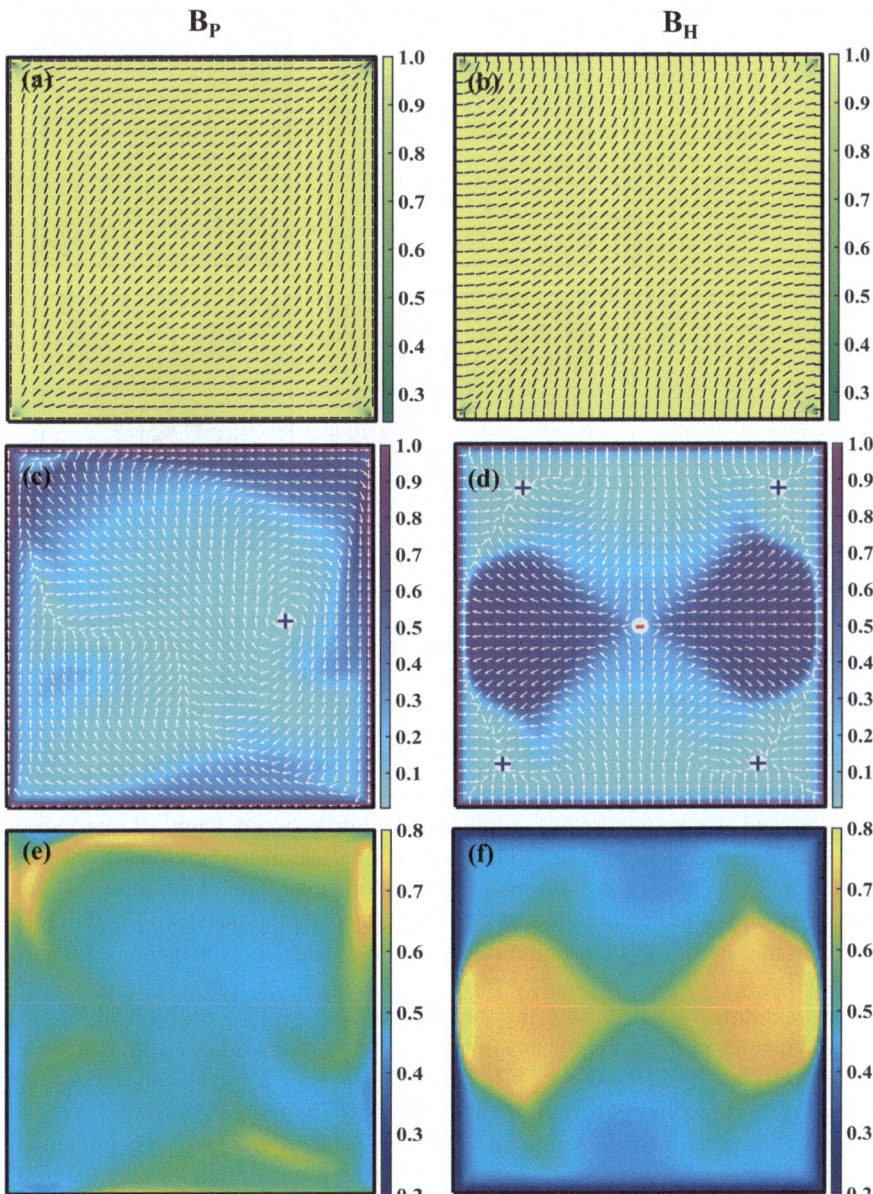

Fig. 5.5 Snapshots at $t = 10^4$ for the **n**-field (first row), **P**-field (second row), and ρ-field (third row) for $c_0 = 0$. The frames **a**, **c** and **e** correspond to planar (B_P) boundary conditions. The color bars in these frames show the nematic orientational order S in (**a**); magnitude of polarization $|P|$ in (**c**); and density ρ in (**e**). The rods (arrows) denote the orientation of the director (polarization) field. The defects are denoted by $+$ or $-$, according to their signs. The corresponding snapshots for homeotropic (B_H) boundaries are shown in frames (**b**), (**d**) and (**f**). Reprinted with permission from [61]. © 2024, American Physical Society. All rights reserved

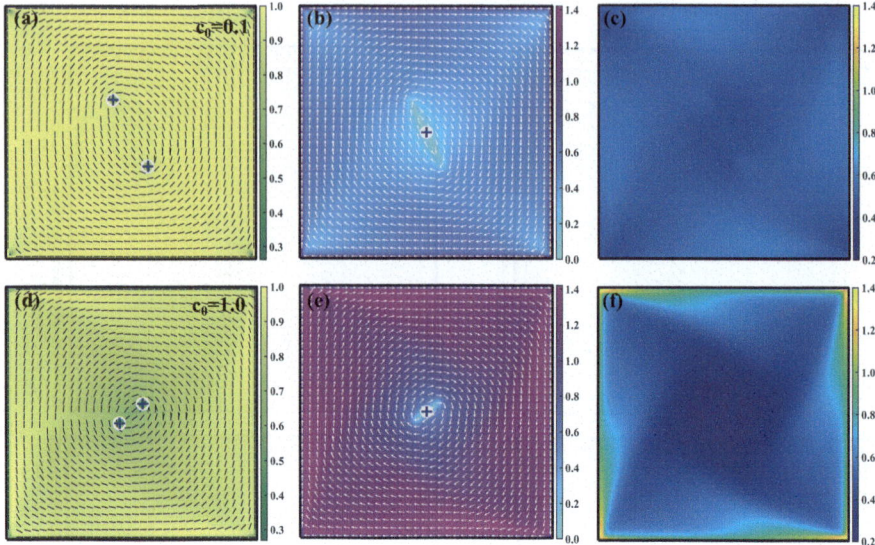

Fig. 5.6 Snapshots at $t = 10^4$ for the coupled case with $c_0 = 0.1$ (upper row) and $c_0 = 1.0$ (lower row). B_P boundary conditions are imposed at the surfaces for both **n** and **P**. The frames **a**, **d** show the **n**-field; **b**, **e** show the **P**-field; and **c**, **f** show the ρ-field. The meaning of various symbols and color bars is the same as in Fig. 5.5. Reprinted with permission from [61]. © 2024, American Physical Society. All rights reserved

Consequently, the observed steady state results from a combined effect of surface anchoring and elastic interactions within the LC. Furthermore, identical topological charges of both nematic and active components produce a structure that consists of two defects with charge $+1/2$. From an application standpoint, these morphologies offer the potential to create pumping effects in microfluidic devices [73]. The sustained circular motion generates fluid flow within the device, which can be directed as desired.

To quantitatively assess how the morphologies in Fig. 5.6 evolve with c_0, the vortex dynamics in the **n**-field is examined. The corotating vortices exhibit stronger binding as c_0 increases, with Fig. 5.7a depicting the time series of the inter-vortex distance d_0 in the asymptotic regime for $c_0 = 0.5, 0.75$, and 1.0.

These time series reveal chaotic fluctuations around an average value, influenced by several factors. First, due to the spatial mesh size $\Delta x = 1$, determining the position of the vortex cores with precision incurs some inaccuracies, especially as c_0 increases and the vortices converge. Furthermore, the square lattice introduces inherent anisotropy because the line connecting the vortex cores is sensitive to alignment with the diagonal of the square well. This contributes to the fluctuations observed in $d_0(t)$ over time. The angular velocity ω_0 of co-rotation over time, shown in Fig. 5.7b for the same c_0-values, indicates similar trends. The log-log plot in Fig. 5.7c of $\bar{d}_0$, the time-averaged $d_0(t)$ in the asymptotic state, versus c_0 suggests that as $c_0 \to 0$,

5.4 Surface-Directed Dynamics in LLCs

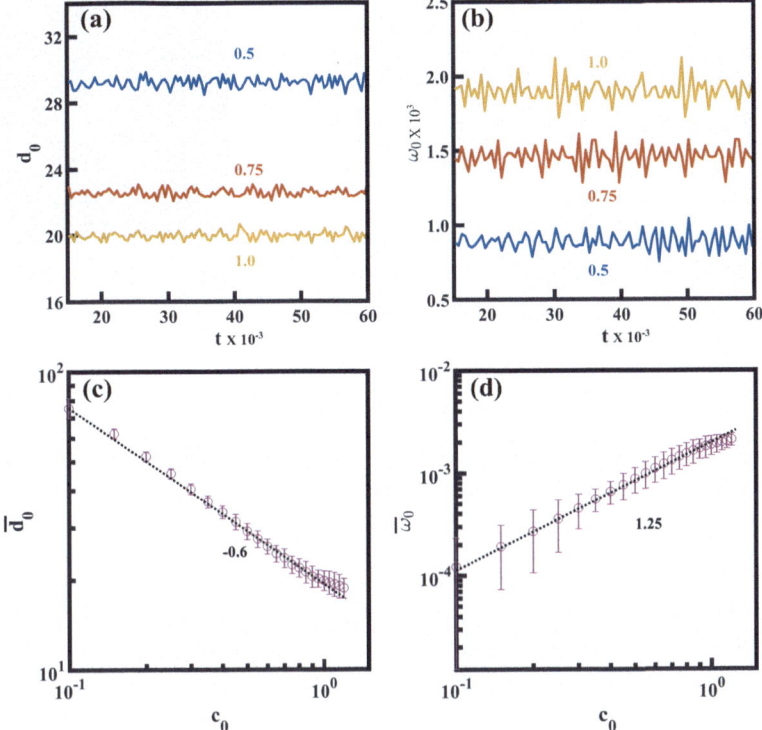

Fig. 5.7 a Plot of the inter-vortex distance d_0 versus t in the asymptotic state for $c_0 = 0.5, 0.75, 1.0$. **b** Plot of the co-rotation angular velocity ω_0 versus t for the same c_0-values. **c** Log-log plot of $\bar{d}_0$ versus c_0. The bar denotes the time-average in the asymptotic state. The dashed line denotes the best linear fit to the data. **d** Log-log plot of $\bar{\omega}_0$ versus c_0. Reprinted with permission from [61]. © 2024, American Physical Society. All rights reserved

$\bar{d}_0$ diverges, consistent with the limit of the uncoupled system. The data is consistent with a power law trend, $\bar{d}_0 \sim c_0^{-\theta}$, where $\theta \approx 0.60$, although saturation occurs beyond $c_0 > 1$. Similarly, Fig. 5.7d presents $\bar{\omega}_0$ versus c_0 on a log-log scale, where in the uncoupled limit ($c_0 \to 0$), $\bar{\omega}_0$ approaches zero. The power-law behavior $\bar{\omega}_0 \sim c_0^\alpha$ with $\alpha \approx 1.25$ holds for $c_0 < 1$. These observations have direct implications for active matter (AM) dynamics. As shown in Fig. 5.6, an increase in c_0 intensifies the swirling effect, with AM components pushed toward the periphery of the well. This behavior results from the interplay between the intrinsic linear velocity of the system v_0 and the coupling-induced angular velocity ω_0. Thus, these results not only illustrate the interdependence between LCs and AM but also outline a method for regulating pattern formation by adjusting the coupling strength.

The second example of coupled kinetics involves a scenario in which the LC exhibits planar boundary conditions (B_P), while the AM is subjected to homeotropic

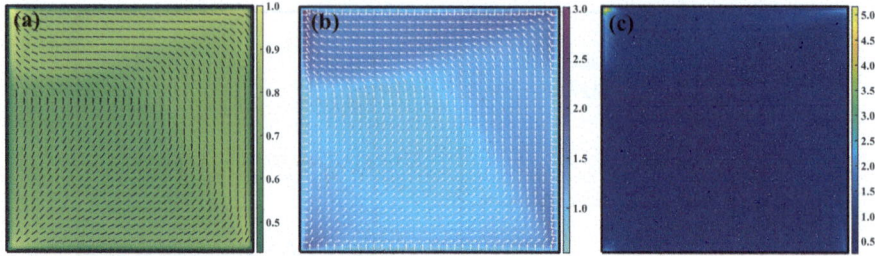

Fig. 5.8 Snapshots at $t = 10^4$ for the coupled case with $c_0 = 1.0$. The boundary conditions for LCs and AM are B_P and B_H, respectively. The frames show the **a** n-field. **b** P-field. **c** ρ-field. The color bars denote the magnitude of the relevant field. Reprinted with permission from [61]. © 2024, American Physical Society. All rights reserved

boundary conditions (B_H). The morphologies resulting from coarsening experiments at a coupling constant $c_0 = 1.0$ are presented in Fig. 5.8.

These snapshots, taken at $t = 10^4$, reflect a system that has reached the FP in its dynamics. In Fig. 5.8a, the orientational order parameter $\mathcal{S}$ is illustrated along with the director orientations for the nematogens. In particular, there are no defects present in the nematic field. The corresponding configurations for the **P**-field and its magnitude are shown in Fig. 5.8b, while the density field ρ is shown in Fig. 5.8c. For comparison, the relevant configurations in the uncoupled limit are illustrated in Fig. 5.5a, d and f. In Fig. 5.5a, FP behavior is evident, while Fig. 5.5d and f showcase complex dynamical states characterized by multiple defects. The results in Fig. 5.8 demonstrate that coupling plays a crucial role in regulating the dynamics of AM, aligning it with the fixed point behavior observed in the LC system. This behavior is reminiscent of the experimental findings reported in [20], where active particles were shown to concentrate in the vicinity of $+1/2$ defects within the nematic field, illustrating the influence of coupling on spatial arrangements in such systems.

5.5 Summary and Discussion

This chapter discusses pattern dynamics in LLCs, which are an amalgamation of AM and NLCs. The proposed model integrates the Toner-Tu framework for AM, the Landau-de Gennes free energy for NLCs, and a coupling term inspired by experimental observations that promotes co-alignment between the polarization in the active medium and nematogens. Unlike early theoretical models limited to dilute regimes without particle interactions, this model includes AM-AM, NLC-NLC, and AM-NLC interactions. This approach captures novel symbiotic dynamics of active and nematic fields. The study focuses on understanding the dynamics of LLCs in $d = 2$. Such systems can be achieved experimentally by confining pure NLCs to shallow wells with planar boundary conditions on the top and bottom surfaces. This

5.5 Summary and Discussion

setup effectively restricts nematic molecules to a single plane with minimal vertical variation.

Our analytical and numerical studies of bulk $d = 2$ LLCs provide a variety of insights, including phase diagrams in different parameter regimes, derived from the fixed point analysis of the proposed model. In particular, the model predicts two unprecedented steady states in LLCs: (i) *chimeras*, which consist of ordered bands of high orientational alignment in both AM and NLC regions, coexisting with disordered regions. These aligned regions in the active and nematic fields move in synchrony at the speed v_0 of the active particles, creating dynamic patterns throughout the system. (ii) *solitons*, localized pockets of order in both active and nematic components, which maintain stability during motion and collisions. While solitons are well-documented in one-dimensional systems, their presence in higher dimensions is rare, adding a unique dimension to the dynamics in LLCs. The resulting induced motion in the passive nematic phase is notably different from the conventional nematic behavior.

We also investigated the influence of BCs on LLC dynamics, with the objective of determining whether specific surface treatments can be engineered to introduce particular dynamic behaviors in LLC systems. This would have potential applications across scientific and technological domains. Two types of BC are analyzed: (a) *planar* or B_P, where the alignment vectors **n** or **P** lie parallel to the surfaces, and (b) *homeotropic* or B_H, where **n** or **P** align perpendicularly to the surfaces. Since $\mathbf{P} \neq -\mathbf{P}$, each BC type has subclasses based on the orientation of **P**. These BCs may arise naturally from confinement within a container or may be intentionally imposed to guide LLC behavior. Two representative configurations of LLCs in square wells are examined.

(a) In the first case, both the **n** and the **P** fields have B_P anchoring, with **P** exhibiting cyclic directionality along the boundaries. In the absence of coupling, the **n**-field remains static while the **P**-field exhibits a single rotating vortex. When coupling is introduced, the system develops a controlled dynamical state characterized by a pair of co-rotating defects in the **n** field. The defect separation and rotational velocity exhibit a power-law relationship with coupling strength.

(b) In the second case, both the fields **n** and **P** are set to B_H, with **P** pointing inward from the surfaces. Without coupling, the **n**-field is static while the **P**-field shows complex, defect-rich dynamics. However, with coupling, this dynamics is subdued, and both **n** and **P** achieve a steady state.

Understanding the interaction between LLCs and their boundary conditions offers avenues for designing active systems with targeted pattern dynamics. The two examples illustrate distinct outcomes: (a) conversion of random motion into a controlled dynamical trajectory, and (b) stabilization of a dynamical configuration into a static state. These cases highlight how surface treatments can be tailored to achieve specific control over LLC dynamics.

In conclusion, our theoretical framework reveals how AM-LC coupling can discipline AM through induced orientational order and restore order in NLCs by removing topological defects. These findings open pathways for the development of self-healing materials capable of targeted information delivery and micro-cargo transport without relying on physical channels. The insights from this study offer multiple approaches to control and manipulate AM and NLCs, with promising potential for innovative applications. The purpose of this work is to inspire further experimental and theoretical explorations in the field of contemporary LLCs.

Appendix: Movies Showing Steady States of LLCs

The movies below show the evolution of the active field (right frame) and nematic field (left frame) to different steady states from the initially disordered state. The steady states exist throughout the simulation time ($t = 50000$), although these are shown in the movies only up to time $t = 2000$.

- Movie 1: Evolution of the LLC into a chimera for *Case 1*. The parameters are $T > T_c$ and $\rho_0 = \rho_c^+ = 0.52$ with the coupling strength $c_0 = 0.5$.
- Movie 2: Evolution of the LLC to the chimera state for *Case 2*: $T < T_c$, $\rho_0 = \rho_c^- = 0.48$ with $c_0 = 0.1$. We point out here that the chimera in the nematic component manifests itself only after the annihilation of all defects (where $S = 0$).
- Movie 3: The 2-dimensional soliton for *Case 2*: $T < T_c$, $\rho_0 = \rho_c^- = 0.48$ with $c_0 = 0.1$. The nematic component exhibits a soliton only after annihilation of all defects.

References

1. E. Ben-Jacob, I. Cohen, O. Shochet, A. Tenenbaum, A. Czirók, T. Vicsek, Phys. Rev. Lett. **75**, 2899 (1995)
2. J.K. Parrish, W.M. Hamner, *Animal Groups in Three Dimensions: How Species Aggregate* (Cambridge University Press, 1997)
3. F.J. Ndlec, T. Surrey, A.C. Maggs, S. Leibler, Nature **389**, 305 (1997)
4. D.K. Helbing, I. Farkas, T. Vicsek, Nature **407**, 487 (2000)
5. D. Helbing, I. Farkas, T. Vicsek, Phys. Rev. Lett. **84**, 1240 (2000)
6. T. Surrey, F. Nédélec, S. Leibler, E. Karsenti, Science **292**, 1167 (2001)
7. S. Hubbard, P. Babak, S. Sigurdsson, K.G. Magnússon, Ecol. Modell. **174**, 359 (2004)
8. A. Sokolov, I.S. Aranson, J.O. Kessler, R.E. Goldstein, Phys. Rev. Lett. **98**, 158102 (2007)
9. V. Schaller, C. Weber, C. Semmrich, E. Frey, A.R. Bausch, Nature **467**, 73 (2010)
10. S. Ramaswamy, Ann. Rev. Cond. Matt. Phys. **1**, 323 (2010)
11. Y. Sumino, K.H. Nagai, Y. Shitaka, D. Tanaka, K. Yoshikawa, H. Chaté, K. Oiwa, Nature **483**, 448 (2012)
12. H.H. Wensink, J. Dunkel, S. Heidenreich, K. Drescher, R.E. Goldstein, H. Löwen, J.M. Yeomans, Proc. Nat. Acad. Sci. **109**, 14308 (2012)
13. J. Palacci, S. Sacanna, A.P. Steinberg, D.J. Pine, P.M. Chaikin, Science **339**, 936 (2013)
14. M.C. Marchetti, J.F. Joanny, S. Ramaswamy, T.B. Liverpool, J. Prost, M. Rao, R.A. Simha, Rev. Mod. Phys. **85**, 1143 (2013)
15. C. Bechinger, R. Di Leonardo, H. Löwen, C. Reichhardt, G. Volpe, G. Volpe, Rev. Mod. Phys. **88**, 045006 (2016)
16. S. Zhou, A. Sokolov, O.D. Lavrentovich, I.S. Aranson, Proc. Nat. Acad. Sci. **111**, 1265 (2014)

References

17. R.R. Trivedi, R. Maeda, N.L. Abbott, S.E. Spagnolie, D.B. Weibel, Soft Matter **11**, 8404 (2015)
18. C. Peng, T. Turiv, Y. Guo, Q. Wei, O.D. Lavrentovich, Science **354**, 882 (2016)
19. J.S. Lintuvuori, A. Würger, K. Stratford, Phys. Rev. Lett. **119**, 068001 (2017)
20. M.M. Genkin, A. Sokolov, O.D. Lavrentovich, I.S. Aranson, Phys. Rev. X **7**, 011029 (2017)
21. A. Sokolov, A. Mozaffari, R. Zhang, J.J. De Pablo, A. Snezhko, Phys. Rev. X **9**, 031014 (2019)
22. S. Zhou, Liq. Cryst. Today **27**, 91 (2018)
23. T. Turiv, R. Koizumi, K. Thijssen, M.M. Genkin, H. Yu, C. Peng, Q.H. Wei, J.M. Yeomans, I.S. Aranson, A. Doostmohammadi, O.D. Laverntovich, Nat. Phys. **16**, 481 (2020)
24. P.G. de Gennes, J. Prost, *The Physics of Liquid Crystals* (Oxford University, Oxford, 1995)
25. M.J. Stephen, J.P. Straley, Rev. Mod. Phys. **46**, 617 (1974)
26. H. Chi, M. Potomkin, L. Zhang, L. Berlyand, I.S. Aranson, Comm. Phys. **3**, 1 (2020)
27. S. Puri, V. Wadhawan, *Kinetics of Phase Transitions* (CRC Press, 2009)
28. A.J. Bray, Adv. Phys. **51**, 481 (2002)
29. J.P. Hernandez-Ortiz, C.G. Stoltz, M.D. Graham, Phys. Rev. Lett. **95**, 204501 (2005)
30. A. Kudrolli, G. Lumay, D. Volfson, L.S. Tsimring, Phys. Rev. Lett. **100**, 058001 (2008)
31. E. Lushi, H. Wioland, R.E. Goldstein, Proc. Nat. Acad. Sci. **111**, 9733 (2014)
32. X. Yang, M.L. Manning, M.C. Marchetti, Soft Matter **10**, 6477 (2014)
33. J. Deseigne, S. Léonard, O. Dauchot, H. Chaté, Soft Matter **8**, 5629 (2012)
34. G.G. Wells, C.V. Brown, Appl. Phys. Lett. **91**, 223506 (2007)
35. R. Barberi, J.J. Bonvent, M. Giocondo, M. Iovane, A.L. Alexe-Ionescu, J. App. Phys. **84**, 1321 (1998)
36. C. Tsakonas, A.J. Davidson, C.V. Brown, N.J. Mottram, Appl. Phys. Lett. **90**, 111913 (2007)
37. H. Kusumaatmaja, A. Majumdar, Soft Matter **11**, 4809 (2015)
38. C. Luo, A. Majumdar, R. Erban, Phys. Rev. E **85**, 061702 (2012)
39. A. Majumdar, A. Lewis, Liquid Crystals **43**, 2332 (2016)
40. M. Robinson, C. Luo, P.E. Farrell, R. Erban, A. Majumdar, Liquid Crystals **44**, 2267 (2017)
41. G. Barbero, L.R. Evangelista, *Adsorption Phenomena and Anchoring Energy in Nematic Liquid Crystals* (CRC Press, 2005)
42. N.J. Mottram, J.P. Newton, (2014). arXiv:1409.3542
43. J. Toner, Y. Tu, Phys. Rev. Lett. **75**, 4326 (1995)
44. J. Toner, Y. Tu, Phys. Rev. E **58**, 4828 (1998)
45. S. Mishra, A. Baskaran, M.C. Marchetti, Phys. Rev. E **81**, 061916 (2010)
46. E. Bertin, M. Droz, G. Grégoire, J. Phys. A **42**, 445001 (2009)
47. T. Ihle, Phys. Rev. E **88**, 040303 (2013)
48. N. Guttenberg, J. Toner, Y. Tu, Phys. Rev. E **89**, 052711 (2014)
49. K. Bisht, V. Banerjee, P. Milewski, A. Majumdar, Phys. Rev. E **100**, 012703 (2019)
50. K. Bisht, Y. Wang, V. Banerjee, A. Majumdar, Phys. Rev. E **101**, 022706 (2020)
51. A. Vats, V. Banerjee, S. Puri, Europhys. Lett. **128**, 66001 (2020)
52. A. Vats, V. Banerjee, S. Puri, Soft Matter **17**, 2659 (2021)
53. A. Vats, S. Puri, V. Banerjee, Phys. Rev. E **106**, 044701 (2022)
54. A. Vats, P.K. Yadav, V. Banerjee, S. Puri, Phys. Rev. E **108**, 024701 (2023)
55. E.W. Cheney, D.R. Kincaid, *Numerical Mathematics and Computing* (Cengage Learning, 2012)
56. Y. Kuramoto, D. Battogtokh, Nonlinear Phenom. Complex Syst. **5**, 380 (2002)
57. D.M. Abrams, S.H. Strogatz, Phys. Rev. Lett. **93**, 174102 (2004)
58. A.C. Newell, *Solitons in Mathematics and Physics* (SIAM, 1985)
59. L.A. Dickey, *Soliton Equations and Hamiltonian Systems* (World Scientific, 1991)
60. S. Puri, Int. J. Mod. Phys. B **4**, 1483 (1990)
61. A. Vats, V. Banerjee, S. Puri, Phys. Rev. E **110**, 034701 (2024)
62. G.P. Alexander, B.G. Chen, E.A. Matsumoto, R.D. Kamien, Rev. Mod. Phys. **84**, 497 (2012)
63. H.H. Wensink, H. Löwen, Phys. Rev. E **78**, 031409 (2008)
64. J. Walton, N.J. Mottram, G. McKay, Phys. Rev. E **97**, 022702 (2018)
65. M. Ravnik, J.M. Yeomans, Phys. Rev. Lett. **110**, 026001 (2013)
66. R. Green, J. Toner, V. Vitelli, Phys. Rev. Fluids **2**, 104201 (2017)
67. H.H. Wensink, H. Löwen, Phys. Rev. E **78**, 031409 (2008)

68. H. Wioland, F.G. Woodhouse, J. Dunkel, J.O. Kessler, R.E. Goldstein, Phys. Rev. Lett. **110**, 268102 (2013)
69. K. Wu, J.B. Hishamunda, D.T. Chen, S.J. DeCamp, Y.W. Chang, A. Fernández-Nieves, S. Fraden, Z. Dogic, Science **355**, 1979 (2017)
70. S. Zhou, O. Tovkach, D. Golovaty, A. Sokolov, I.S. Aranson, O.D. Lavrentovich, New J. Phys. **19**, 055006 (2017)
71. S.M. Fielding, D. Marenduzzo, M.E. Cates, Phys. Rev. E **83**, 041910 (2011)
72. Y. Fily, A. Baskaran, M.F. Hagan, Soft Matter **10**, 5609 (2014)
73. S.P. Thampi, A. Doostmohammadi, T.N. Shendruk, R. Golestanian, J.M. Yeomans, Sci. Adv. **2**, e1501854 (2016)

Chapter 6
Conclusion and Perspectives

Abstract The main aspects of our theoretical framework to study pattern dynamics for two-component soft matter systems are discussed. The versatility of the phenomenological approach and interesting future applications of this formalism are presented in a cohesive manner.

In this research monograph, we have introduced and studied coarse-grained models for *nematic liquid crystals* (NLCs) with two classes of inclusions: *magnetic nano particles* (MNPs) and *active matter* (AM). The composite systems are known as *ferronematics* (FNs) and *living liquid crystals* (LLCs), respectively, and have attracted considerable recent experimental attention. The dynamics of these complex systems is studied to explore unique morphologies, defect configurations, and novel phases that originate from the interplay of the two components. This concluding chapter provides an overview of the proposed theoretical framework and potential directions for future studies.

Before proceeding, it is useful to examine the utility and limitations of coarse-grained or macroscopic models. Typically, physical systems can be studied at either the microscopic or macroscopic levels of description. Microscopic-level models solve the equations of motion for molecules with "realistic" interactions between them. These models are relatively easy to formulate and closer to experimental reality. However, because of their discrete nature, they are not amenable to theoretical analysis. Further, due to computational constraints, they are only useful for accessing physical behavior at short length scales and early times. On the other hand, coarse-grained models consist of partial differential equations for appropriate order parameters, as we have seen in the preceding chapters. The relevant free energy functionals are constructed from symmetry considerations, which may not always be straightforward. Because the order parameters are coarse-grained entities, these models are not applicable at molecular length and time scales. Rather, they are useful for understanding phase diagrams, morphologies, defects, and scaling behavior. Furthermore, finer details such as particle shape, size, and density can be integrated, enhancing the accuracy of predictions and agreement with experiment.

Let us return to the specific systems we have discussed in this book. The advent of FNs has opened avenues for studying novel magnetic, optical, and mechanical properties. Although significant experimental progress has been made in this direction, studies of pattern dynamics in FNs remain scarce. This monograph has addressed the theoretical gap by using kinetic models based on the well-established Landau-de Gennes free energy framework. These studies show that the two-component coupling can give rise to *slaved coarsening*, where nematic or magnetic ordering occurs even above their critical temperatures. A systematic control over defect configurations is possible through adjustments in the coupling strength and quench temperatures. Another key observation is the emergence of *biaxiality* in $d = 2$ FNs, which is absent in pure $d = 2$ NLCs.

The framework developed here is versatile and can be easily adapted for other NLC-based systems. For example, the recently discovered *ferroelectric* system, which exhibits unique blue phases in experimental studies [1, 2], can be analyzed using this approach. Similarly, single-component systems, such as bent core NLCs, which exhibit both polar and orientational order [3], can also be explored. In addition to the above systems, the effect of quenched/annealed disorder on the dynamics of FNs is a subject of ongoing experimental and theoretical interest.

We have also extended the above phenomenological modeling to the study of LLCs. In this case, the system is intrinsically nonequilibrium due to the presence of AM. We perform detailed simulations of this model that show novel steady-state solutions, such as *chimeras* and *solitons*. The phase diagram of our model can be obtained from an analysis of the fixed points and their linear stability analysis. The model developed here includes a complete range of interactions between AM-AM, NLC-NLC, and AM-NLC. Further insights emerge from confining these suspensions in shallow wells as a result of the interplay between boundary conditions and activity. Our framework demonstrates that AM-NLC coupling can control active matter by inducing orientational order and heal NLCs by eliminating topological defects. The findings have significant implications, especially for the design and synthesis of new self-healing materials, and constitute a significant step toward harnessing flows in AM.

The proposed LLC model offers several directions for future development. Our current approach includes a coupling term that enforces parallel alignment between the nematic and active components. However, experimental observations in dilute systems reveal scenarios in which nematic alignment directs active motion, while active particles simultaneously disrupt the nematic order. In such a situation, the framework would require non-reciprocal interaction terms in the governing equations. Furthermore, the present model neglects hydrodynamic effects in the nematic medium, focusing solely on phenomena driven by a free energy plus advection. This assumption is suitable for dense suspensions, where hydrodynamic fields decay rapidly. However, it excludes key features of the system, such as flow-driven instabilities, dynamic defect-defect interactions, and nonlinear responses to external forces. Inclusion of hydrodynamic flows would enable the study of turbulence, defect-mediated flow, and other rich behaviors, significantly broadening the scope of LLC dynamics.

In this book, the effect of confinement on LLCs has been studied in $d = 2$. This is appropriate for thin-film geometries, a setup often employed in LLC experiments. However, when dealing with systems of finite height, it becomes essential to extend the analysis to $d = 3$ to account for molecular orientations along the vertical axis. Theoretical studies in $d = 3$ are expected to reveal even more interesting phenomena. A notable difference in the $d = 3$ LdG free energy is the presence of the term $\text{Tr}(\mathbf{Q})^3$, which allows first-order transitions in the nematic component.

The framework developed in the preceding chapters emphasizes the impact of inclusions on the dynamics of NLCs. It also sets the stage for studying pattern formation in multi-component soft matter systems from distinct physical settings. Some examples include polymer mixtures, colloids, and biological materials [4–7]. We hope that this book has laid the foundation for future studies that will enable new discoveries of fundamental importance and technological relevance.

References

1. N. Sebastián, M. Čopič, A. Mertelj, Phys. Rev. E **106**, 021001 (2022)
2. O.D. Lavrentovich, Proc. Nat. Acad. Sci. **117**, 14629 (2020)
3. A. Jákli, O.D. Lavrentovich, J.V. Selinger, Rev. Mod. Phys. **90**, 045004 (2018)
4. S.R. Nagel, Rev. Mod. Phys. **89**, 025002 (2017)
5. H. Wu, H. Friedrich, J.P. Ptterson, N. Sommerdijk, N. De Jonge, Adv. Mater. **32**, 2001582 (2020)
6. L.S. Hirst, *Fundamentals of Soft Matter Science* (CRC Press, 2019)
7. A. Fernandez-Nieves, A.M. Puertas, *Fluids, Colloids and Soft Materials: An Introduction to Soft Matter Physics* (Wiley, 2016)

Appendix A
Stationary Solutions

In this Appendix, we present analytical results for stable fixed points of our TDGL model in Eqs. (3.6)–(3.9) for FNs in $d = 2$. We consider all the cases in Table 3.1 (Tables A.1, A.2 and A.3).

Table A.1 Coupling limits and stable stationary solutions for Case 1 of Table 3.1. This corresponds to a quench at temperature T such that $T_c^N < T < T_c^M$

Coupling limits	Stable stationary solutions $(Q_{11}^*, Q_{12}^*, M_1^*, M_2^*)$
(i) $c_1 \neq 0$, $c_2 = 0$	$(r_Q, 0, 1, 0)$ $r_Q = c_1(1 + \hat{S})^{-1}$ $\hat{S} = S^2/4 = 3^{-1}(-2 + a_1 + a_1^{-1})$ $a_1 = 2^{1/3}(2 + c_1^2 + c_1\sqrt{54 + 272 + c_1^2})^{-1/3}$
(ii) $c_1 = 0$, $c_2 \neq 0$	$(0, 0, 1, 0)$
(iii) $c_1 = c_2 = c$	$(r_Q, 0, r_M, 0)$ $r_Q = c\,(1 + \hat{S} + c^2)^{-1}$ $\hat{S} = S^2/4 = (1 + c^2)(a_1 - 2/3) + a_1^{-1}$ $3a_1 = 2^{1/3}(35c^2 + 6c^4 + 2c^6 + a_2^{1/2})^{-1/3}$ $a_2 = 1053c^4 + 54c^2(2c^2 + 6c^4 + 2c^6)$ $r_M = \left[(1 + \hat{S})(1 + c^2 + \hat{S})^{-1}\right]^{1/2}$

Table A.2 Coupling limits and stable stationary solutions for Case 2 of Table 3.1. This corresponds to a quench at temperature T such that $T_c^M < T < T_c^N$

Coupling limits	Stable stationary solutions $(Q_{11}^*, Q_{12}^*, M_1^*, M_2^*)$
(i) $c_1 \neq 0, c_2 = 0$	$(1, 0, 0, 0)$
(ii) $c_1 = 0, c_2 \neq 0$	$(1, 0, r_M, 0)$ $r_M = \sqrt{c_2 - 1}$
(iii) $c_1 = c_2 = c$	$(r_Q, 0, r_M, 0)$ $r_Q = c\,(1 - \hat{S} + c^2)^{-1}$; $\hat{S} = S^2/4 = 0.5(2 + c^2 + \sqrt{4c^2 + c^4})$ $r_M = \left[(\hat{S} - 1)(1 + c^2 - \hat{S})^{-1}\right]^{1/2}$

Table A.3 Coupling limits and stable stationary solutions for Case 3 of Table 3.1. This corresponds to a quench at temperature T such that $T < \min\{T_c^M, T_c^N\}$

Coupling limits	Stable stationary solutions $(Q_{11}^*, Q_{12}^*, M_1^*, M_2^*)$
(i) $c_1 \neq 0, c_2 = 0$	$(r_Q, 0, 1, 0)$ $r_Q = -c_1(1 - \hat{S})^{-1}$; $\hat{S} = S^2/4 = 1/3(-2 + a_1 + a_1^{-1})$ $a_1 = 2^{1/3}(a_2 + c_1\sqrt{-54 + 27a_2})^{-1/3}$
(ii) $c_1 = 0, c_2 \neq 0$	$(1, 0, r_M, 0)$ $r_M = \sqrt{c_2 + 1}$
(iii) $c_1 = c_2 = c$	$(r_Q, 0, r_M, 0)$ $r_Q = c\,(1 - \hat{S} + c^2)^{-1}$ $\hat{S} = S^2/4 = 0.5(2 + c^2 + \sqrt{4c^2 + c^4})$ $r_M = \left[(1 - \hat{S})(1 + c^2 - \hat{S})^{-1}\right]^{1/2}$

The manufacturer's authorised representative in the EU is Springer Nature Customer Service Centre GmbH, Europaplatz 3, 69115 Heidelberg, Germany. If you have any concerns regarding our products, please contact ProductSafety@springernature.com

Printed and bound by CPI Group (UK) Ltd, Croydon, CR0 4YY

26/03/2026

02078977-0002